T0291629

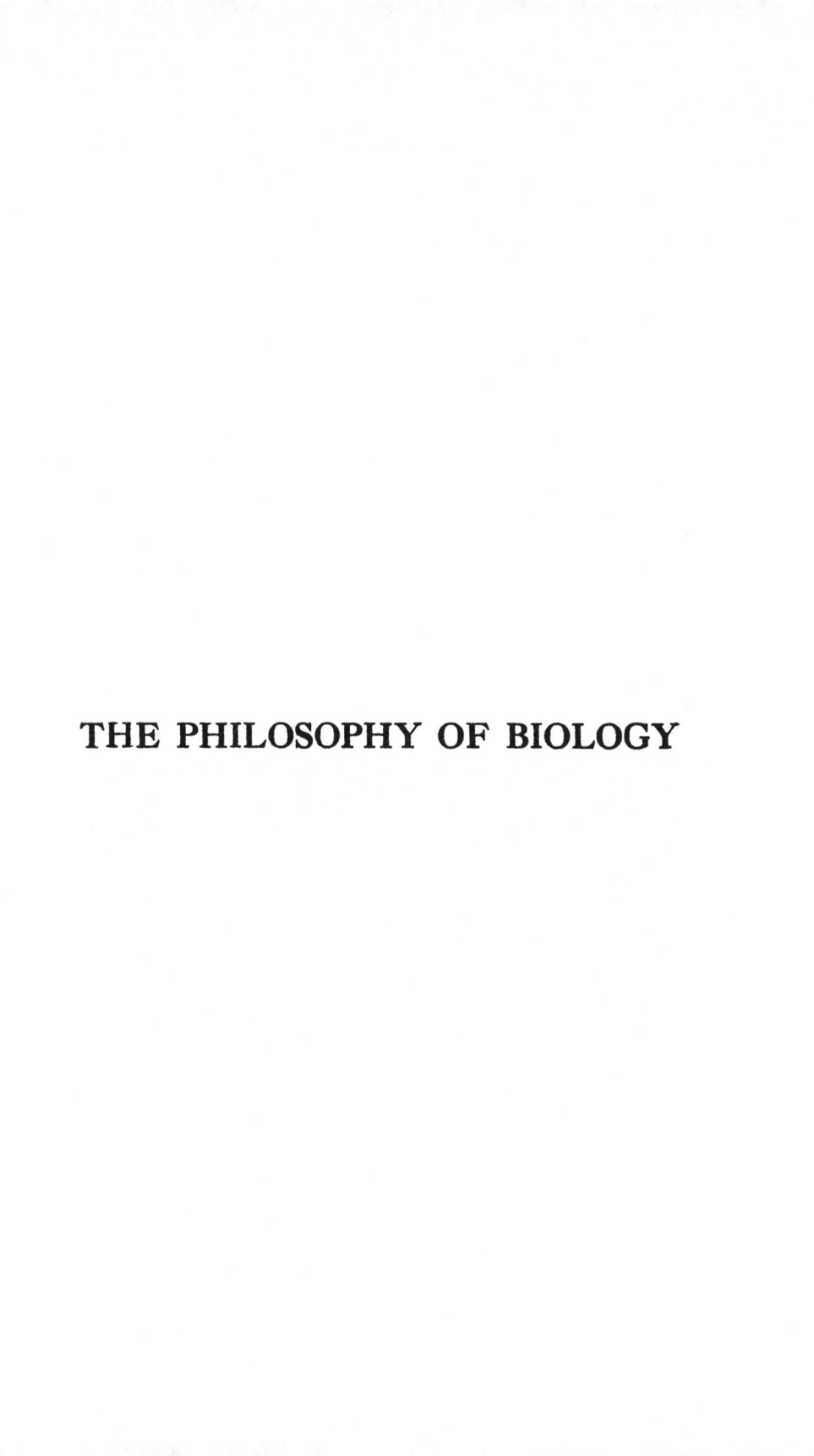

THE PHILOSOPHY OF BIOLOGY

THE PHILOSOPHY

OF

BIOLOGY

BY

JAMES JOHNSTONE, D.Sc.

Cambridge:

at the University Press

1914

CAMBRIDGE
UNIVERSITY PRESS

University Printing House, Cambridge CB2 8BS, United Kingdom

Published in the United States of America by Cambridge University Press, New York

Cambridge University Press is part of the University of Cambridge.

It furthers the University's mission by disseminating knowledge in the pursuit of education, learning and research at the highest international levels of excellence.

www.cambridge.org
Information on this title: www.cambridge.org/9781107644137

© Cambridge University Press 1914

This publication is in copyright. Subject to statutory exception and to the provisions of relevant collective licensing agreements, no reproduction of any part may take place without the written permission of Cambridge University Press.

First published 1914
First paperback edition 2014

A catalogue record for this publication is available from the British Library

ISBN 978-1-107-64413-7 Paperback

Cambridge University Press has no responsibility for the persistence or accuracy of URLs for external or third-party internet websites referred to in this publication, and does not guarantee that any content on such websites is, or will remain, accurate or appropriate.

INTRODUCTION

IT has been suggested that some reference, of an apologetic nature, to the title of this book may be desirable, so I wish to point out that it can really be justified. Science, says Driesch, is the attempt to describe Givenness, and Philosophy is the attempt to understand it. It is our task, as investigators of nature, to describe what seems to us to happen there, and the knowledge that we so attain—that is, our perceptions, thinned out, so to speak, modified by our mental organisation, related to each other, classified and remembered—constitutes our Givenness. This is only a description of what seems to us to be nature. But few of us remain content with it, and the impulse to go beyond our mere descriptions is at times an irresistible one. Fettered by our habits of thought, and by the limitations of sensation, we seem to look out into the dark and to see only the shadows of things. Then we attempt to turn round in order that we might discover what it is that casts the shadows, and what it is in ourselves that gives shape to them. We seek for the Reality that we feel is behind the shadows. That is Philosophy.

The Physics of a generation earlier than our own thought that it had discovered Reality in its conception of an Universe consisting of atoms and molecules in ceaseless motion. What it described were only motions and transformations, but it understood these motions and transformations as matter and energy. Yet more subtle minds than the great physicists of the beginning of the nineteenth century had already seen that sensation might mislead us. There was something in us that continually changed—that was our consciousness, and it was all that we knew. If external things did exist they existed only because we thought them. But we ourselves exist, for we are not only a stream of consciousness that continually changes, but there is in us a personality, or identity, which has remained the same throughout all the vicissitudes of our consciousness. If the things that exist for us exist only because we think them, and if we also exist, then we must exist in the thought of an Absolute Mind that thinks us. Physical Science, studying only motions and transformations, understood that there was something that moved and transformed—this was matter and energy. Mental Science, studying only thought, understood that nature was only the thought of an Universal Mind. Either conclusion was equally valid Philosophy (or metaphysics), and neither could be proved or disproved by the methods of Science. The speculative game is drawn, said Huxley, let us get to practical work !

Both Physics and Biology did get to work, with the results that we know. But Physics advanced far beyond the acquirement of the results that stimulated Biology to formulate our present hypotheses of evolution and heredity. As its knowledge accumulated, it began to doubt whether matter and energy, atoms and molecules, mass and inertia—all those things which it thought at first were so real—were anything else after all than ways in which our mental organisation dealt with crude sensations. They might, as Bergson said later on, be the moulds into which we pour our perceptions. Physics set up a test of Reality, the law of the conservation of matter and energy. There are existences which may or may not persist. Visions and phantasms and dreams are existences *while they last*. They are true for the mind in which they occur. But they seem to arise out of nothing, and to disappear into nothing, and physical Science cannot investigate them. They are existences which are not conserved. On the other hand those images which we call moving matter and transforming energy can be investigated by the methods of physics. Molecules change, but something in them, the atoms, remain constant. Energy becomes transformed, and it may even seem to cease to exist, but if it disappears, then something is changed so that the lost energy can be traced in the nature of the change. Matter and energy are conserved and therefore they are the only Realities. But the test is obviously one that has an *a priori* basis, and we may doubt whether it is a test of Reality.

Thus Physics constructed a dynamical Universe, that is, one which consisted of atoms which attracted or repelled each other with forces which were functions of the distances between them. Even now this conception of a dynamical, Newtonian Universe is a useful one, though we recognise that it is only symbolism. But it was not a conception with which Physics could long remain content. How could atoms separated from each other by empty space act on each other, that is, how could a thing act where it was not ? There must be something between the atoms. The Universe could not be a discontinuous one, and so Physics invented an Universe that was *full*. It was an immaterial, homogeneous, imponderable, continuous Universe. That which existed behind the appearances of atoms and molecules and energy was the ether of space. It must be admitted that the conception appears to the layman to involve only contradictions : heterogeneous, discontinuous, ponderable atoms are only singularities in a homogeneous, continuous, imponderable medium, or ether. Yet it is easy to see that this contradiction arises in our mind only because we had previously thought of the Universe in terms of matter and energy, and in spite of ourselves we attempt to think of the new Reality in terms of the old one. In its attempt to understand all its later results Physics had therefore to invent a new Philosophy—that of the ether of space.

It is only in our own times that Biology has become sceptical and has begun to doubt whether its earlier Philosophy is a sound one. That which it describes— the object-matter of its Science—is not that which Physics describes. There are two domains of Givenness, the organic and the inorganic. Biology, leaning on Physics, studied motions and transformations, just as Physics did, though the motions which it studied were more complex and the transformations more mysterious. But borrowing the methods of investigation of Physics it borrowed also its Philosophy, and so it placed behind its Givenness the Reality that Physics at first postulated and then abandoned. The organism was therefore a material system actuated by energy. The notion, it should be noted, is not a deduction from the results of Biology, but only from its methods.

Did Physiology, that is, the Physiology of the Schools, ever really investigate the organism ? A muscle-nerve preparation, an excised kidney through which blood is perfused, an exposed salivary gland which is stimulated, even a frog deprived of its cerebral hemispheres—these things are not organisms. They are not permanent centres of action, autonomous physico-chemical constellations capable of independent existence, and capable of indefinite growth by dissociation. They are parts of the organism, which, having received the impulse of life, an impulse which soon becomes exhausted, exhibit for a time some of the phenomena of the organism. What Physiology did attain in such investigations was an analytical description of some of the activities of the organism. It did not describe life, but rather the physico-chemical reactions in which life is manifested. The description, it should be noted, is all-important for the human race in its effort to acquire mastery over its environment ; and there is no other way in which it may be carried further but by the methods of physical Science. Givenness is one, though we arbitrarily divide it into the domains of the organic and the inorganic, and there can be only one way of describing it. That is the mechanistic method.

Nevertheless all this is only a description, and our Philosophy must be the attempt to understand our description. The mechanistic biologist, in the attempt to identify his Philosophy with that of a former generation of physicists, says that he is describing a physico-chemical aggregate—an assemblage of molecules of a high degree of complexity—actuated by energy, and undergoing transformations. But our scepticism as to the validity of this conclusion is aroused by reflecting on its origin. If it was borrowed from the Philosophy of a past Physics, and if the more penetrating analysis of the Physics of our own time has made a new Philosophy desirable, should not Biology also revise its understanding of its descriptions ? For Biology has not stood still any more than Physics, and the Physiology of our own day has become different from that of the times when the mechanistic Philosophy of life took origin. The embryologists and the naturalists of our own generation have studied the *whole* organism in its normal functioning and behaviour, and have obtained results which cannot easily be understood as physico-chemical mechanism. Life is not the activities of the organism, but the integration of the activities of the organism, just as Reality for Physics is not the atoms and molecules of gross matter, but the integration of these in the ether of space.

This, then, is all that we mean by the philosophy of Biology—the attempt to understand the descriptions of the Science in the light of its later investigations. Philosophy, in the academic sense, we have not considered in relation to the subject-matter of our science, though there is much in the classic systems that is of absorbing interest, even to the working investigator of the nineteenth century. The biological education is not, however, such as to predispose one towards these studies. The reader will recognise that the point of view, and the methods of treatment, adopted in this book are those suggested by Driesch and Bergson, even if no references are given. He may, perhaps, appreciate this limitation ; for, influenced by the modern scientific training, he may be inclined to regard Philosophy as Mark Twain regarded his Egyptian mummy : if he is to have a corpse it might as well be a real fresh one.

J. J.

LIVERPOOL
November 1913

CONTENTS

CHAPTER I

Argument.—The conscious organism is one that acts. Its consciousness of an external world is not simply the result of the stimuli made by that world on its organs of sense, for it becomes fully aware only of those stimuli which result in deliberated bodily activity. This awareness of an outer world on which it acts is the perception of the organism. Its consciousness is an intensive multiplicity. This multiplicity is arbitrarily dissociated, for convenience' sake, by the mental organisation, which confers extension and magnitude and succession on those aspects of consciousness which it arbitrarily dissociates from each other. Our notion of space is an intuitive one and depends on our modes of bodily exertion. Our notions of motion and continuity are also intuitive ones, and they cannot be represented intellectually, but we can approximate to them by the methods of the infinitesimal calculus. Mathematical time is only a series of standard events which 'punctuate our duration. Duration is the accumulated existence and experience of the organism. We cannot prove intellectually that there is a world external to our consciousness, but that this world exists is a conviction intuitively held.

CHAPTER II

Argument.—If the organism is a physico-chemical mechanism its activities must conform to the two principles of energetics : the law of conservation of energy and matter, and the law of entropy-increase. They conform strictly to the law of conservation. The law of the degradation of energy is true of our experience of inorganic nature, but we can show that it cannot be universally true. Inorganic processes are irreversible ones, and they proceed in one direction only, and in them energy is degraded. Organic processes, that is, the processes carried on in the generalised organism, are irreversible; or, at least, there is a tendency for them to be carried on without necessary dissipation of energy.

CHAPTER III

Argument.—If the organism is investigated by the methods of physical and chemical science, nothing but physico-chemical activities can be discovered. This is necessarily the case, since methods which yield physico-chemical results only are employed. The physiologist makes an analysis of the activities of the organism, and he reduces these activities to certain categories; although all attempts completely to describe the functioning of the organism solely in terms of physical and chemical reactions fail. In addition to the reactions which make up the functioning of an organ or organ-system, there is direction and co-ordination of these reactions. The individual physico-chemical reactions which occur in the functioning of the organism are integrated, and life is not merely these reactions, but also their integration.

CHAPTER IV

Argument.—The notion of the organism as a physico-chemical mechanism is a deduction from the methods of physiology, and not from its results. The notion of vitalism is a natural or intuitive one. The historic systems of vitalism assumed the existence of a spiritual agency in the organism, or of a form of energy which was peculiar to the activities of the organism. Modern investigation lends no support to either belief. But the study of the organism as a whole, that is, the study of developmental processes, or that of the organism acting as a whole, afford a logical disproof of pure mechanism. It shows that there cannot be a functionality, in the mathematical sense, between the inorganic agencies that affect the whole organism and the behaviour or functioning of the whole organism. Mechanism is only suggested in the study of isolated parts of the organism. We are compelled toward the belief that there is an agency operative in the activities of the organism which does not operate in purely inorganic becoming. This is the Vital Impetus of Bergson, or the Entelechy of Driesch.

CHAPTER V

Argument.—The concept of the organic individual is one which is arbitrary, and is convenient only for purposes of description. Life on the earth is integrally one. Personality is the intuition of the conscious organism that it is a centre of action, and that all the rest of the universe is relative to it. The individual organism, regarded objectively, is an isolated, autonomous constellation, capable of indefinite growth by dissociation, differentiation, and re-integration. This growth is reproduction. The dissociated part reproduces the form and manner of functioning of the individual organism from which it has proceeded. The offspring varies from the parent organism, but it resembles it much more than it varies from it. There are therefore categories of organisms in nature the individuals of which resemble each other more than they resemble the individuals belonging to other categories : these are the elementary species. Hypotheses of heredity are corpuscular ones, and are based on the physical analogy of molecules and atoms. The concept of the species is a logical one. The organism is a phase in an evolutionary or a developmental flux, and the idea of the species is attained by arresting this flux.

CHAPTER VI

Argument.—A reasoned classification of organisms suggests that a process of evolution has taken place. It suggests logical relationships between organisms, while the results of embryology and palæontology suggest chronological relationships. Yet this kinship of organisms might only be a logical, and not a material one. Evolution may have occurred somewhere, but it might be argued that the ideas of species have generated each other in a Creative Thought. But transformism may be produced experimentally, and so science has adopted a mechanistic hypothesis of the nature of the process. Transformism of species depends on the occurrence of variations, but these arise spontaneously and independently of each other, and they must be co-ordinated. This co-ordination of variations cannot be the work of the environment. Variations are cumulative, and they exhibit direction, and this direction is either an accidental one, or it is the expression of an impetus or directing agency in the varying organism itself. The problem of the cause of variation is only a pseudo-problem.

CONTENTS

CHAPTER VII

Argument.—If we assume the existence of an evolutionary process, the results of morphology, embryology, and palæontology ought to enable us to trace the directions followed during this process. But these results are still so uncertain that they indicate only a few main lines of transformism. Phylogenetic trees are largely conjectural in matters of detail. Evolution has resulted in the establishment of several dominant groups of organisms—the metatrophic bacteria, the chlorophyllian organisms, the arthropods, and the vertebrates. Each of these groups displays certain characters of morphology, energy-transformation, and behaviour ; and a certain combination of characters is concentrated in each of the groups. But there is a community of character in all organisms which have arisen during the evolutionary process. The transformation of kinetic into potential energy is characteristic of the chlorophyllian organisms. The utilisation of potential energy, and its conversion into the kinetic energy of regulated bodily activity, by means of a sensori-motor system, is characteristic of the animal. The bacteria carry to the limit the energy-transformations begun in the tissues of the plants and animals. Immobility and unconsciousness characterise the plant, mobility and consciousness the animal. Animals indicate two types of actions— intelligent actions and instinctive actions. Instinctive activity involves the habitual exercise of modes of action that have been inherited. Intelligent activities involve the exercise of modes of action that are not inherited, but which are acquired by the animal during its own lifetime, and are the results of perceptions which show the animal that its activity is relative to an outer environment.

CHAPTER VIII

Argument.—A strictly mechanistic hypothesis of evolution compels us to regard the organic world, and the inorganic environment with which it interacts, as a physico-chemical system. All the stages of an evolutionary process must therefore be equally complex : they are simply phases, or rearrangements, of the elements of a transforming system. The physics on which these mechanistic hypotheses were based was that of a discontinuous, granular, Newtonian universe, that is, one consisting of discrete particles, or mass-points, attracting or repelling each other with forces which are functions of the distances between them.   It was a spatially extended system of parts.   Therefore at all stages in an evolutionary process, or one of individual development, the elements of the system constitute an extensive manifoldness, and the obligation of mechanistic hypotheses of evolution and development to accept this view has shaped modern theories of heredity. Life is an intensive manifoldness, but in individual or racial evolution this intensive manifoldness becomes an extensive manifoldness.   Life is a bundle of tendencies which can co-exist, but which cannot all be fully manifested, in the same material constellation, therefore these tendencies become dissociated in the evolutionary process.   In this dissociation there is direction and co-ordination, which are the Vital Impetus of Bergson, or the Entelechy of Driesch.

Entelechy is an elemental agency in nature which we are compelled to postulate because of the failure of mechanism.   It is not spirit, nor a form of energy, but the direction and co-ordination of energies.   There is a sign, or direction of inorganic happening which absolutely characterises the processes which are capable of analysis by physico-chemical methods of investigation, and the result of this direction of inorganic happening is material inertia. Yet this direction cannot be universal : it must be evaded some· where in the universe.   It is evaded by the organism.

The problem of the nature of life is only a pseudo-problem.

APPENDIX

Infinity and the notion of the limit.   Functionality.   Frequency distributions and probability.   Matter, force, mass, and inertia. Energy-transformations.   Isothermal and adiabetic transformations.   The Carnot engine and cycle.   Entropy.   Inert matter.

THE PHILOSOPHY OF BIOLOGY

CHAPTER I

THE CONCEPTUAL WORLD

LET us suppose that we are walking along a street in a busy town ; that we are familiar with it, and all the things that are usually to be seen in it, so that our attention is not likely to be arrested by anything unusual ; and let us further suppose that we are thinking about something interesting but not intellectually difficult. In these circumstances all the sights of the town, and all the turmoil of the traffic fail to impress us, though we are, in a vague sort of way, conscious of it all. Electric trams approach and recede with a grinding noise ; a taxicab passes and we hear the throb of the engine and the hooting of the horn, and smell the burnt oil ; a hansom comes down the street and we hear the rhythmic tread of the horse's feet and the jingle of the bells ; we pass a florist's shop and become aware of the colour of the flowers and of their odour ; in a café a band is playing " ragtime." There are policemen, hawkers, idlers, ladies with gaily coloured dresses and hats, newsboys, a crowd of people of many characteristics. It is all a flux of experience of which we are generally conscious without analysis or attention, and it is a flux which is never for a moment quite the same, for everything in it melts and flows into everything else. The noise of the tram-cars is incessant, but now and then it becomes

A

louder ; the music of the orchestra steals imperceptibly on our ears and as imperceptibly fades away ; the smell of the flowers lingers after we pass the shop, and we do not notice just when we cease to be conscious of it ; the rhythm of the ragtime continues to irritate after we have ceased to hear the band—all the sense-impressions that we receive melt and flow over into each other and constitute our stream of consciousness, and this changes from moment to moment without gap or discontinuity. It is not a condition of " pure sensation," but it is as nearly such as we can experience in our adult intellectual life.

It is easy to discover that many things must have occurred in the street which did not affect our full consciousness. We may learn afterwards that we have passed several friends without recognising them ; we may read in the newspapers about things that happened that we might have seen, but which we did not see ; we may think we know the street fairly well, but we find that we have difficulty in recalling the names of three contiguous shops in it ; if we happen to see a photograph which was taken at the time we passed through the street we are usually surprised to find that there were many things there that we did not see. Why is it, then, that so much that might have been perceived by us was not really perceived ? We cannot doubt that everything that came into the visual fields of our eyes must have affected the terminations of the optic nerves in the retinas ; the complex disturbances of the air in the street must have set our tympanic membranes in motion ; and all the odoriferous particles inhaled into our nostrils must have stimulated the olfactory mucous membranes. In all these cases the stimulation of the receptor organs must have initiated nervous impulses, and these must have been propagated along the sensory nerves, and must have reached the brain, affecting masses of nerve cells there. Nothing in physiology seems to indicate that we can inhibit or repress the activity of the distance sense-receptors, visual, auditory, and olfactory, with their central connections in the brain ; they must have functioned, and must have been physically affected by the events that took place outside ourselves, and yet we were unconscious, in the fullest sense of this term, of all this activity. Why is it, then, that our perception was so much less than the actual physical reception of external stimuli that we must postulate as having occurred ? Sherlock Holmes would have said that we really saw and heard all these things although we did not observe them, but the full explanation involves a much more careful consideration of the phenomena of perception than this saying indicates.

There is, of course, no doubt that we did see and hear and smell all the things that occurred in the street during our aimless peregrination, that is, all the things which so happened that they were capable of affecting our organs of sense. This is true if we mean by seeing and hearing and smelling merely the stimulation of the nerve-endings of the visual, auditory, and olfactory organs, and the conduction into the brain of the nervous impulses so set up. But merely to be stimulated is only a part of the full activity of the brain ; the stimulus transmitted from the receptor organs must result in some kind of bodily activity if it is to affect our stream of consciousness. Two main kinds of activity are induced by the stimulation of a receptor organ and a central ganglion, (1) those which we call reflex actions, and (2) those actions which we recognise as resulting from deliberation. We must now consider what are the processes that are involved in these kinds of neuro-muscular activity.

The term " reflex action " is one that denotes rather a scheme of sensori-motor activity than anything that actually happens in the animal body ; it is a concept that is useful as a means of analysis of complex phenomena. In a reflex three things happen, (1) the stimulation of a receptor organ and of the nerve connecting this with the brain, (2) the reflection, or shunting, of the nervous impulse so initiated from the *terminus ad quem* of the afferent or sensory nerve, to the *terminus a quo* of the efferent or motor nerve, and (3) the stimulation of some effector organ, say a motor organ or muscle, by the nervous impulse so set up. The simplest case, perhaps, of a reflex is the rapid closure of the eyelids when something, say a few drops of water, is flicked into the face. Stated in the way we have stated it the simple reflex does not exist. In the first place, it is a concept based on the structural analysis of the complex animal where the body is differentiated to form tissues—receptor organs, nerves, muscles, glands, and so on. But a protozoan animal, a *Paramœcium* for instance, responds to an external stimulus by some kind of bodily activity, and yet it is a homogeneous, or nearly homogeneous, piece of protoplasm, and this simple protoplasm acts at the same time as receptor organ, conducting tissue or nerve, and effector organ. In the higher animal certain parts of the integument are differentiated so as to form visual organs, and the threshold of these for light stimuli is raised while it is lowered for other kinds of physical stimuli. Similarly other parts of the in-- tegument are modified for the reception of auditory stimuli, becoming more susceptible for these but less susceptible for other kinds of stimuli than the adjacent parts of the body. Within the body itself certain tracts of protoplasm are differentiated so that they can conduct molecular disturbances set up in the receptor organs in the integument better than can the general protoplasm; these are the nerves. Other parts are modified so that they can contract or secrete the more easily; these are the muscles and glands. The conception of a reflex action, as it is usually stated in books on physiology, therefore includes this idea of the differentiation of the tissues, but all the processes that are included in the typical reflex are processes which can be carried on by undifferentiated protoplasm.

It is also a schematic description that assumes a simplicity that does not really exist. As a rule a reflex is initiated by the stimulation of more than one receptor organ, and the impulses initiated may thus reach the central nervous system by more than one path. There is no simple shunting of the afferent impulse from the cell in which it terminates into another nerve, when it becomes an efferent impulse; but, instead of this, the impulse may " zigzag " through a maze of paths in the brain or spinal cord connecting together afferent and efferent nerves and ganglia. Further, the final part of the reflex, the muscular contraction, is far from being a simple thing, for usually a series of muscles are stimulated to contract, each of them at the right time and with the right amount of force, and every contraction of a muscle is accompanied by the relaxation of the antagonistic muscle. There are muscles which open the eyelids and others which close them, and the cerebral impulse which causes the levators to contract at the same time causes the depressors to relax.

It is quite necessary to remember that the simple

reflex is really a process of much complexity and may involve many other parts and structures than those to which we immediately direct our attention. But leaving aside these qualifications we may usefully consider the general characters of the reflex, regarding it as a common, automatically performed, restricted bodily action, involving receptor organ, central nervous organ, and effector organ. There are certain kinds of external stimuli that continually affect our organs of sense, and there are certain kinds of muscular and glandular activity that occur " as a matter of course," when these stimuli fall on our organs of sense. The emanation from onions or the vapour of ammonia causes our eyes to water ; the smell of savoury food causes a flow of saliva ; and anything that approaches the face very rapidly causes us to close the eyes. Reflexes are, in a way, commonly occurring, purposeful and useful actions, and their object is the maintenance of a normal condition of bodily functioning.

We dare hardly say that the simple reflex is an unconsciously performed action, although we are not conscious, in the fullest sense of the term, of the reflexes that habitually take place in ourselves. But even in the decapitated frog, which moves its limbs when a drop of acid is placed on its back, something, it has been said, akin to consciousness may flash out and light up the automatic activity of the spinal cord. We must not think of consciousness as that state of acute mentality which we experience in the performance of some difficult task, or in some keenly appreciated pleasure, or in some condition of mental or bodily distress ; it is also that dimly felt condition of normality that accompanies the satisfactory functioning of the parts of the bodily organism. But this dim and obscure feeling of the awareness of our actions is easily inhibited whenever what we call intellectual activity proceeds.

Much of the stimulation of our receptor organs is of this generally occurring nature, and we are not aware of it although the stimuli received are such as to induce useful and purposeful bodily activity. In walking along the street we automatically avoid the people, and the other obstacles that we encounter, by means of regulated movements of the body and limbs, but this is activity that has become so habitual and easy that we are hardly aware of it, and not at all, perhaps, of the physical stimuli which induce it. But not only do we receive stimuli which are reflected into bodily actions without our being keenly aware of this reception, but we also receive stimuli which do not become reflected into bodily activity. It is, Bergson suggests, as if we were to look out into the street through a sheet of glass held perpendicularly to our line of sight ; held in this way we see perfectly all that happens in front of us, but when we incline the glass at a certain angle it becomes a perfect reflector and throws back again the rays of light that it receives. This is, of course, a physical analogy, and no comparison of material things with psychical processes can go very far, but in a way it is more than an analogy. In our indolent absorbed state of mind we do not as a rule see the objects which we are not compelled to avoid, and which do not, in any way, influence our immediate condition of bodily activity. The optical images of all these things are thrown upon our retinas and are, in some way, thrown or projected upon the central ganglia, but there the series of events comes to an end, for the images are not reflected out towards the periphery of the body as muscular actions. We cannot doubt that this is why we do not perceive all the stimulation of our organs of sense that we are sure that take place. These stimuli pass through us, as it were, unless they are reflected out again as actions. In this reflection, or translation of neutral into muscular activity, perceptions arise.

But even then perception need not arise. It does not, as a rule, accompany the automatically performed reflex action, because the latter is the result of intra-cerebral activities that have become so habitual that they proceed *without friction*. There are innumerable paths in the brain along which impulses from the receptor organs may pass into the motor ganglia, but in the habitually performed reflex actions these paths have been worn smooth, so to speak. The images of objects which are perceived over and over again by the receptor organs glide easily through the brain and as easily translate themselves into muscular, or some other kind of activity. The things that matter in the life of an animal which lives " according to nature " are cyclically recurrent events in which, after a time, there is nothing new. Most of them proceed just as well in the animal deprived of its cerebral hemispheres by operation as in the intact cerebrate animal. In the performance of actions of this kind the organism becomes very much of an automaton.

Let something unusual happen in the street while we are walking through it—a runaway horse, or the fall of an overhead " live " wire, for instance, something that has seldom or never formed part of our experience, and something that may have an immediate effect on us as living organisms. Then perception arises at once because the stimulation of our organs of sense presents us with something which is unfamiliar, and yet not so unfamiliar that it does not recall from memory, or from derived experience, reminiscences of the images of somewhat similar things, and of the effects of these. The train of events that now proceeds in our central nervous system becomes radically different from that which proceeded in our former, rather aimless, series of actions. The stimuli no longer pass easily through the " lower " ganglia of the brain, but flash upwards into the cortical regions, where they become confronted with the possibility of innumerable alternative paths and connections with all the parts of the body. They waver, so to speak, before adopting one or other, or a combination of these paths; there is hesitation, deliberation, and finally choice of a path, with the result that a series of muscular organs become inervated and motor actions, of a type more or less competent to the situation in which we find ourselves, are set up. In this hesitation and deliberation perception arises. It is when the animal may act in a certain way as the result of a stimulus which is not a continually recurrent one, but at the same time may refrain from acting, or may act in one of several different ways, that perception of external things and their relations arises.

That is to say, we perceive and think because we act. We do not look out on the environment in which we are placed in a speculative kind of way, merely receiving the images of things, and classifying and remembering them, while all the time we are passive in so far as our bodily activities are concerned. If the results of modern physiology teach us anything in an unequivocal way they teach us this—that the organs of activity, muscles, glands, and so on, and the organs of sense and communication, are integrally one series of parts, and that apart from motor activity nervous activity is an aimless kind of thing. It is because we act that we think and disentangle the images of things presented to us by our organs of sense, and subject all that is in the stream of consciousness to conceptual analysis.[1]

That is to say, in thinking about the flux of consciousness we decompose it into what we regard as its constituent parts, and we confer upon these parts separate existence in space and time. But it is clear that none of the things which we thus regard as the elements of our consciousness has any real existence apart from the others. The smell of the flowers and that of the burnt oil interpenetrate in our consciousness of the stimulation of our olfactory organs just as do the jingle of the cab bells, the music of the orchestra, and the throb of the motor car in the impressions transmitted by our auditory organs. It is difficult to see that all these things, with the multitude of other things which we perceive, constitute a " multiplicity in unity," that is an assemblage of things which are separate things, but which do not lie alongside each other in space and mutually exclude each other, but which are all jammed into each other, so to speak. It is easy to see that we are conscious of a heterogeneity, and whenever we think of this multitude of things it seems natural that we should separate them from each other. The stream of our consciousness is so complex that we cannot attend to it all at once, not even to the few things that we have picked out in our example. If we concentrate our attention on any part, or rather aspect of it, all the rest ceases to exist, or rather we agree to ignore it, and this very concentration of thought upon one part of our experience isolates it from all the rest. To a certain extent the analysis of the complex of sensation is the result of the work of different receptor organs ; certain

---

[1] All this is, of course, the argument of Bergson's earlier books, *Matière et Mémoire* and *Données immédiates de la Conscience*.

fields of energy, which we call light, radiation, etc., affect the nerve-endings in the retina ; chemically active particles in the atmosphere affect the nerve-endings in the olfactory membranes ; and rapidly repeated changes of pressure in the atmosphere (sound vibrations) affect the auditory organs in the internal ear, and so on. But this reception of different stimuli by different receptor organs exists only in the higher animal; there are no specialised sense organs in a *Paramœcium*, for instance, and the whole periphery of the animal must receive all these different kinds of external stimuli at once. The specialisation of its receptor organs in the higher animal is rather the means whereby the organism becomes more receptive of its environment, than the means whereby it analyses that environment. This analysis is the work of the consciousness of the animal.

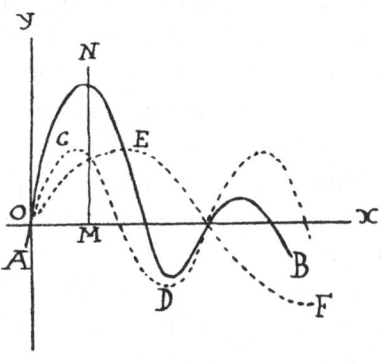

FIG. I.

Suppose that we draw a curve *AB* freehand with a single undivided sweep of the pencil. By making a certain assumption—that the curve which we drew was one that might be regarded as cyclical, that is, might be repeated over and over again—we can subject it to harmonic analysis. We can decompose it into a number of other curves (*CD*, *EF*, etc.), each of which is a separate " wave " rising above and falling below the axis *OX* in a symmetrical manner. If we draw any vertical line *MN* cutting these curves, we shall find that the distance between the axis *OX* and the main curve AB is always equal to the algebraic sum of the distances between the axis and the other curves. These latter we call the harmonic constituents of the curve AB, supposing them to " add up " so as to form it. But AB was something quite simple and elemental and its constituents cannot be said to have existed in it when we drew it freehand ; it was only by an artifice of practical utility in mathematical computations that we *constructed* them. It may be, of course, that the harmonic constituents of a curve had actual existence apart from the curve itself, but, in the case that we take, they certainly had not. Now we must think of our stream of consciousness in much the same way. It is something immediately experienced and elementary; it is the concomitant, if we choose so to regard it, of the external processes that go on outside our bodies. We can investigate it by thinking about it, and attending to one aspect of it after another, thus arbitrarily detaching one " part " of it from all the rest, but immediately we do this we rise above the flux of experience into the region of intellectual concepts. We have converted a multiplicity of states of consciousness, all of which co-exist along with each other, and in each other, and which have no spatial existence, into a multiplicity of states, visual, auditory, olfactory, etc., which have become separated from each other and have therefore acquired extension. This dissociation of the flux of experience is the process of conceptual analysis carried out by thought.

If we dissociate the stream of consciousness in this way, breaking it up into states which we choose to regard as separate from each other, we shall see that of the elements which we thus isolate many are like each other and can be associated. Obviously there is a greater resemblance between different smells than between smells and sounds. Different musical sounds are more like each other than are sounds, and feelings of heat and cold. There is a greater likeness between the states of consciousness which arise from the stimulation of the same receptor organ, than between those that arise from the stimulation of different receptors. Those differences of sensation accompanying the stimulation of different sense organs we regard as different in kind ; there is absolutely no resemblance between a colour and a sound, we say, however much the modern annotator of concert programmes may suggest the analogy. But we say that there may be different degrees of stimulation of the same sense organ, and that the sensations that we thus receive are of the same kind though they differ in intensity. The whistle of a railway engine becomes louder as the train approaches, that is to say, more intense, and if we study the physical conditions that are concomitant with the stimulation of our tympanic membranes we shall see that waves of alternate rarefaction and compression are set up in the atmosphere outside our ears. All the time that the train approaches the frequency of these waves remains the same, that is, just as many occur in a second when the train is distant as when it is near. But the amplitude of the waves has been increasing, and the velocity with which the molecules of air strike against the tympanic membranes becomes greater the nearer is the source of sound. We can represent this by means of a diagram which shows that the amplitude of the waves— which represents the loudness of the sound—increases while the frequency—which represents the pitch— remains the same. The amplitude is represented by the straight vertical lines, 11, 22, 33, etc., which are of increasing magnitude. Thus we represent the physical cause of the increasing loudness of the sound by space-magnitudes, and then we transfer these magnitudes to the states of consciousness concomitant with the vibrating molecules of air. Suppose that we knew nothing at all about the cause of the differences of pitch of musical sounds and that we listen to the notes of the octave, C, D, E,——C, sounded by an organ ; all that we should experience would be that the sounds were different. If we were to sing the notes we might attain the intuition that the notes G, A, B were "higher" than the notes C, D, E, because a greater effort was required in order to produce these sounds, but obviously this is a different thing from saying that

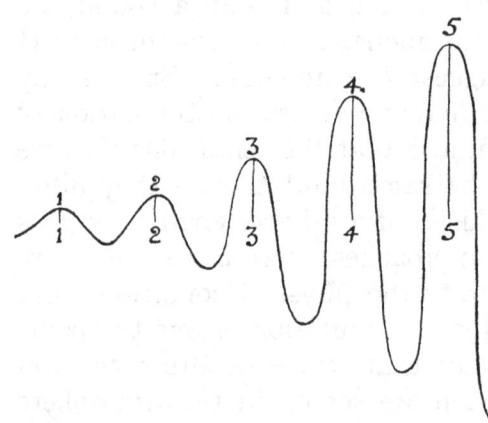

Fig. 2.

the notes themselves were "higher" or "lower." But let us match the notes by striking tuning-forks, and then having selected forks which give the notes of the octave let us fix them so that they will make a tracing, while still vibrating, on a revolving strip of paper. We shall then find that the fork emitting the note C makes (say) 256 vibrations per second, the fork D $\frac{9}{8}$ 256 vibrations, the fork E $\frac{5}{4}$ 256 vibrations, and so on. Thus we associate the notes of the octave together and we say that their quality was the same but that their pitch differed, and since the pitch depends on the frequency of vibration of the fork, or of the air in its vicinity, we say that pitch differences are quantitative ones, and that the states of consciousness which accompany these physical events are also quantitatively different.

So also with colour. If we had no such apparatus as prisms or diffraction gratings, which enable us to find what is the wave length of light, should we have any idea of the spectral hues, red, yellow, orange, green, etc., as differing from each other quantitatively? It is certain that we should not. But observation and experiment have shown that the nerve-endings of the optic nerve in the retina are stimulated by vibrations of something which we agree to call the ether of space, and that the frequency of vibration of light which we call red is less than that which we call orange, while the frequency of vibration of orange light is less again than that of blue light, and so on. To our consciousness red, orange, yellow, and blue light are absolutely different, but we disregard this intuition and we say that our perceptions of light are similar in kind but differ, in some of them are more intense than are some others. Again, have we any intuitive knowledge of increasing temperature? If we dip our hands into ice-cold water the sensation is one of pain, if the water has a temperature of 5° C. it feels cold, if it is at 15° C. we have no particular appreciation of temperature, if at 25° C. it feels very warm, if it is at 60° it is very hot, and if it is at 90° we are probably scalded and the feeling is again one of pain. If we place a thermometer in the water we notice that each sensation in turn is associated with a progressive lengthening of the mercury thread, and if we investigate the physical condition of the water we find that at each stage the velocity of movement of the molecules was greater than that at the preceding stage. We say, then, that our different perceptions were those of heat of different degrees of intensity, so transferring to the perceptions themselves the notions of space-magnitudes acquired by a study of the expansion of the mercury in the thermometer, or by the adoption of the physical theory of the kinetic structure of the water.   Yet it is quite certain that what we experienced were quite different things or conditions, cold, warmth, heat, and pain, and indeed, in this series of perceptions different receptor organs are involved.

Suppose we listen to the note emitted by a syren which is sounding with slowly increasing loudness but with a pitch which remains constant.   We do not notice at first that the sound is becoming louder, but after a little time we do notice a difference.   Let us call the amplitude of vibration of the air when the syren first sounds E, and then, when we notice a difference, let us call the amplitude $\Delta E + E$, ΔE being the increment of amplitude.   Let us call our sensation when the syren first sounds S, and our sensations when the sound has become louder $S + \Delta S$, ΔS being the " increment of sensation."   Then the relation holds :—

$$\frac{\Delta E}{E} = \text{constant.}$$

That is to say, the louder is the sound the greater must be the increase of loudness before we notice a difference. Let us assume now that the successive sensations of loudness that we receive as the syren blows louder and louder are, each of them, just the same amount louder than the preceding sound ; that is to say, let us assume that what we experience are " minimal perceptible differences " of sensation—that they are " elements of loudness "—thus we construct a series of sounds each of which differs from that preceding it by an elemental increment of loudness.   Now things that cannot be further decomposed are necessarily equal to each other ; if, for instance, the atoms represent the ultimate units into which we break up the matter called oxygen, then these atoms are all equal to each other. Therefore the increments of loudness are equal to each other.

If we plot these equal increments of loudness as the dependent variable S in a graph, and the amplitude of the vibrations of the atmosphere as the independent variable E, we can obtain the following curve   If we investigate this we shall find that a certain relation exists between the " values " of the sensation and the values of the stimuli that correspond to them ; a regular increase in the loudness of the sensation corresponds to a regular increase in the log-

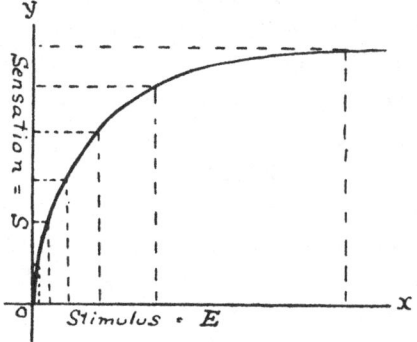

Fig. 3.

arithms of the strength of the stimuli. Let $S =$ the sensation, E the stimulus, and C and Q constants ; then

$$S = C \log \frac{E}{Q} \, ;$$

so that we seem to establish a mathematical relation between the intensity of our sensations and the intensity of the stimuli that give rise to those sensations, but this relation depends on the assumption that what we call " minimal perceptible differences " of sensation are numerical differences that are equal to each other, and this is, of course, an assumption that cannot possibly be proved.

B

Thus we decompose our stream of consciousness into a series of quantitatively different and qualitatively different things, upon each of which we confer independent existence. We attribute to these different aspects of our consciousness extension, but the extension is due only to our analysis ; for the qualities of pitch, loudness, colour, odour, etc., which we disentangle from each other, did not exist apart from each other, any more than do the sine and cosine curves into which we decompose an arbitrarily drawn curved line. The multiplicity of our consciousness is intensive, like the multiplicity that we see to exist in the abstract number ten. This number stands for a group of things, but its multiplicity is intensive and only exists because we are able to subdivide anything in thought to an indefinite extent. Now, so far we have only separated what we agree to regard as the elemental parts of our general perception of the environment, but it is to be noted that we have not given to these elements anything like spatial extension.

We may, if we like, regard our intuition of space as that of an indefinitely large, homogeneous, empty medium which surrounds us and in which we may, in imagination, place things. So regarded it is difficult to see in what way our notion of space differs from our idea of " nothing," a pseudo-idea incapable of analysis, except into the idea of something which might be somewhere else. The more we think about it the more we shall become convinced that space, that is the " form " of space, represents our actual or potential modes of motion, that is, our powers of exertional activity. Space, we say, has three dimensions ; in all our analysis of the universe, and of the activities that we can perceive in it, this idea of movement in three dimensions, forward and backward, up and down, and right and left, occurs; and we have to recognise that in it there is something fundamental, as fundamental as the intuitive knowledge that we possess of the direction of right and left. It is because we can move in such a way that any of our motions, no matter how complex, can be resolved into the components of backward and forward, right and left, and up and down, these directions all being at right angles to each other, that we speak of our movements as three-dimensional ones. Our geometry is founded, therefore, on concepts derived from our modes of activity; and there is nothing in the universe, apart from our own activity, that makes this the only geometry possible to us. Euclidean geometry does not depend on the constitution of the external universe, but on the nature of the organism itself.

There is a little Infusorian which lives, in its adult phase, on the surface of the spherical ova of fishes. These ova float freely in sea water, and the Infusorian crawls on their surfaces, moving about by means of ciliary appendages. It does not swim about in the water, but adheres closely to the surface of the ovum on which it lives. Let us suppose that it is an intelligent animal and that it is able to construct a geometry of its own; if so, this geometry would be very different from our own.

It would be a two-dimensional geometry, for the animal can move backward and forward, and right and left, but not up and down; it is a stereotropic organism, as Jacques Loeb would say, that is, it is *compelled* by its organisation to apply its body closely to the surface on which it lives. But its two-dimensional geometry would, on this account, be different from ours. Our straight lines are really the *directions* in which we move from one point to another point in such a way as to involve the least exertion ; they are the shortest distances between two points, and if we deviate from them we exert a greater degree of activity than if we had moved along them. For us there is only one straight line that can be drawn between two points, but this is not necessarily true for our Infusorian, and its straight line need not be the shortest distance between two points. It might be either the longest or the shortest distance between the points, for the latter can always be placed on a great circle passing through the two points and the poles of the egg, and in moving from a point on which it is placed the animal could reach the other point by moving in two directions, just as we could go round the earth along the equator by moving to the east or to the west. Therefore the straight line of the Infusorian would be not only a scalar quantity but a vector quantity, that is, it would represent, not only a quantity of energy, but a quantity of energy that has direction. For us only one straight line can be drawn between two given points, but this limitation would not exist in the two-dimensional geometry of a curved surface. Suppose that the two points are situated on a great circle and that they are exactly 180° apart ; then the Infusorian could move from one pole to another pole along an infinite number of straight lines or meridians all of which had a different direction, but all of which were of the same length ; that is to say, in this geometry an infinite number of straight lines can be drawn between the same two points. Again, its triangles *might* be different from ours ; our triangles are figures formed by drawing straight lines between three points, and on a plane surface the sum of the angles of the triangle are together equal to two right angles, though on a curved surface they may be greater or less than two right angles. But our Infusorian could not imagine a triangle in which the sum of the angles was not greater than two right angles, for all its figures would be drawn on a convex surface.

Our three-dimensional geometry depends, therefore, on our modes of activity and the concepts with which it operates; points, straight lines, etc. are conceptual limits to those modes of activity. We can imagine a straight line only as a direction along which we can move without deviating to the right or the left, or up or down. But even if we draw such a line on paper with a fine pencil the trace would still have some width, and we can imagine ourselves small enough to be able to deviate to the right or the left within the width of the line drawn on the paper. We might make a very small mark on the paper, but no matter how small this mark is it would still have some magnitude; otherwise we should be unable to see it. If the straight line had no width and the point no magnitude they would have no perceptual existence. Our perceptual triangles are not figures, the angles of which are necessarily equal to two right angles. If we drive three walking sticks into a field and then measure the angles between them by means of a sextant we shall find that the sum is *nearly* 180°, but in general not that amount. If we stick a darning needle into the heads of each of the walking sticks and then remeasure the angles by means of a theodolite we shall obtain values which are nearer to that of two right angles, but we should not, except by " accident," obtain exactly this value. We do not, therefore, get the " theoretical " result, and we say this is because of the errors of our methods of observation; but why do we suppose that there is such a theoretical result from which our observations deviate, if our observations themselves do not in general give this ideal result ? We might accumulate a great series of measurements of the angles of our triangle, and we should then find that these results would tend to group themselves symmetrically round a certain value which would be 180°. Some of the results would be considerably less than the ideal, and some of them would be considerably more ; but these relatively great deviations would be small in number and most of the results would be a very little less than 180° or a very little more, and there would be as many which would be a little less as those that were a little more. We should have formed a " frequency distribution "[1] with its " mode " at 180°.

But by " reasoning " about the " properties " of these lines and triangles in plane two-dimensional space, we should arrive at the conclusion that the angles of a triangle were equal to 180°, and neither more nor less. We should then think of a straight line as still a *path* along which we move in imagination, and a path which still has some width. But we imagine the width of the path to become less and less, so that, even if we imagine ourselves to become thinner and thinner, we should be unable to deviate either to the right or left in moving along the path, because the thinner we make ourselves the thinner becomes also the path. We imagine our intuition of a deviation to the right or left becoming keener and keener, so that, no matter how small the deviation we should still be able to appreciate it by the extra exertion which it would involve. We think of a point as a little spot, and we think of ourselves as being very small indeed, so that we can move about on this spot. But we can reduce the area of the spot more and more, until it

---

[1] See appendix, p. 350.

becomes " infinitesimally " small ; and at the same time we think of ourselves as becoming smaller and smaller, so that we can still move about on the spot. But we think of the area of the spot as becoming so small that no matter how small we make ourselves we are unable to move on it.

This means that we substitute conceptual lines and points and triangles for the perceptual ones of our experience, and then we operate in imagination with these concepts. That is to say, we carry our modes of exertional activity to their *limits*,[1] in the way which we have tried to indicate above—a process of thought which is the foundation of the reasoning of the infinitesimal calculus.

What we call space, therefore, depends on our intuition of bodily exertion. This intuition includes the knowledge that a certain change has occurred as the consequence of the expenditure of a certain amount of bodily energy, and that, as the result of this change, the relation of the rest of the universe to our body has become different. We think of our body as the origin, or centre, of a system of co-ordinates :—

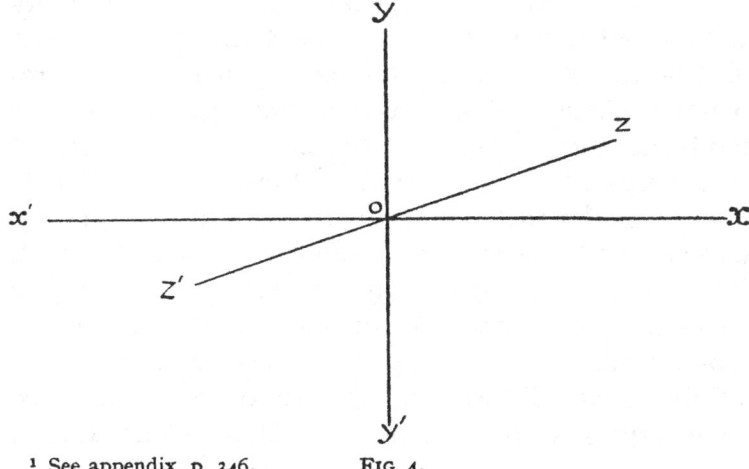

We imagine three lines at right angles to each other to extend indefinitely out into space, and we think of ourselves as being situated at the point of intersection of these three straight lines. If anything moves in the universe outside ourselves we can resolve this motion into three components, each of which is to be measured along one of the axes of our system of co-ordinates. But any motion whatever in the universe outside ourselves can be represented equally well by supposing that the origin of the system of co-ordinates has been changed ; that is, by supposing that *we* have changed our position relative to the rest of the universe. Therefore motion outside ourselves is not to be distinguished from a contrary motion of our own body—a statement of the " principle of relativity "— except that any change outside ourselves may be distinguished from that compensatory change in the position of our body which *appears* to be the same thing, by the absence of the intuition that we have expended a certain quantity of energy in producing the change. Conscious motion of our own body is something *absolute* ; all other motion is relative.

So far we have been speaking of our crude bodily motion, but a very little consideration will show that our knowledge of space attained by scientific measurements depends just as much on our intuition of our bodily activity, and its direction ; the measurement of a stellar parallax, or that of the meridian altitude of the sun, for instance, by astronomical instruments, involves bodily exertion, though of a refined kind. Three-dimensional space, that is *our* space, therefore represents the manner of our activity, just as convex two-dimensional space represents the manner of the activity of the Infusorian, and one-dimensional space would represent the manner of activity of an animal which was compelled to live in a tube, the sides of which it fitted closely, so that it could move only in one direction—up and down. A parasite, living attached to some fixed object, and the movements of which were represented only by the growth of its tissues, could not form any idea of space ; and the " higher " forms of geometry, that is, space of four or more dimensions, present no clear notion to our minds, even although we regard the operations included in mathematics of this kind as pure symbolism, because we cannot relate this imaginary space to any form of bodily exertion. Geometry, then, represents the manner in which our bodily exertion cuts up the homogeneous medium in which we live.

Motion, whether it be that of our own body in controlled muscular activity, or that imaginary motion of the environment which we call giddiness, or a sensibly perceived motion of some part of the environment, that is, a motion which we can compensate by some actual or imaginary change in the position of our own body produced by our own exertion, is an intuitively felt change, and is incapable of intellectual representation. It is not clearly conceived either in ancient or in modern geometry. Euclidean geometry is, as we have seen, based directly on our intuition of bodily exertion, but it is essentially static in treatment. Let it be admitted that we can draw a straight line of any length and in any direction, and so on ; then we regard these straight lines, etc., as motionless, abstract things, and we proceed to discuss their relationships. Cartesian geometry, and the methods of the infinitesimal calculus, do not treat of real motion, and the concept, if it is introduced at all, is introduced illegitimately and surreptitiously. Consider what we do when we " plot a curve." Let the latter be a parabola having the equation $y=\frac{1}{2} x$. Now a parabola is defined as " the locus of a point which *moves*, so that its distance from a fixed point is in a constant relation to its distance from a fixed straight line." How do we construct such a curve ?

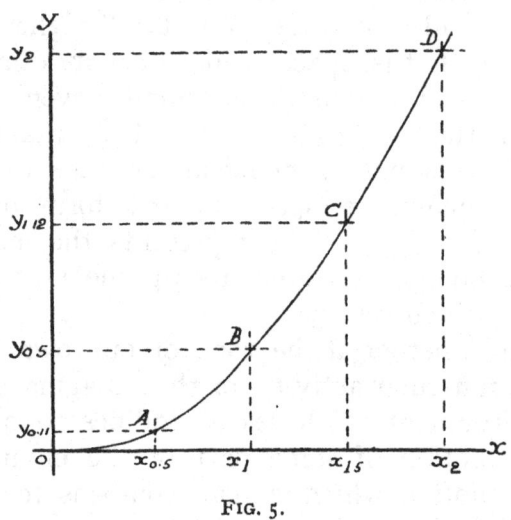

FIG. 5.

We proceed to fix the positions of a series of points in this way: there are two straight lines, OX and OY, at right angles to each other, and we measure off certain steps along the line OX; these steps are $OX_{0.5}$, OX_1, $OX_{1.5}$, OX_2, and so on, the small numerals indicating the distance of each point ($OX_{0.5}$, etc.) from the origin O. We then draw lines perpendicular to the X-axis through these points. We have now to calculate one-half of the square of each of these lengths $OX_{0.5}$, OX_1, etc., and then we mark off these calculated lengths along the perpendicular lines. The point A, for instance, is $\frac{1}{2}(0.5)^2$ from the point $X_{0.5}$, B is $\frac{1}{2}(1)^2$ from X_1, and so on. In this way we obtain a series of points, A, B, C, D, E, etc., and these are points on the locus of the " moving " point.

There is nothing at all about motion here. All that we have done is to measure lengths. We have made a kind of counterpoint, X-points against Y-points, but we have not even made a curve. We connect the points A, B, C, D, E, etc., by means of short, straight lines, and then we may connect together these short lines, and, if we plot a number of intermediate points between those that we have already obtained and join these, the points may be so close together that they may seem to be indistinguishable from a curve. Yet, no matter how numerous they may be, they can never be connected together so as to form a curve ; we therefore draw a curved line freehand through them, and at once, in so doing, we abandon our intellectual methods, for our curve depends on our intuition of *continuously changing direction*.

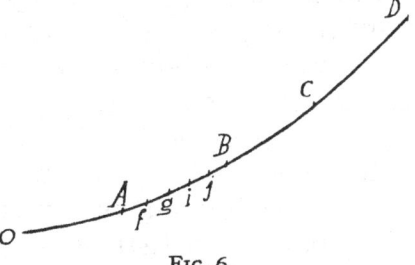

Fig. 6.

But if we think about it we shall find that we can form no clear intellectual notion of continuity and we can only measure the curvature of a line *at a point in* the line by drawing a tangent to the curve at this point, and then by measuring the slope of the tangent. The curve itself we obviously leave out of consideration.

We cannot conceive of the point moving along the locus OD. We can think of it only as *at* the places O, A, B, C, D, E, etc., but we must neglect the intervals OA, AB, BC, CD, DE, and so on, or we can divide them into smaller intervals by supposing the point to have occupied the positions f, g, i, j, between the points A and B, for instance. Yet, no matter how many these intervals may be, we can only think of the point as being *at* the places O, A, B, C, D, E, or at f, g, i, j, and so on. We never think of the intervals themselves, and, if all we think about is the *position* of the point, we do not really think of it as in motion at all. We can *see* it in motion, but we cannot form an intellectual concept of its motion. It is not really necessary that we should in the affairs of everyday life, but for the adequate treatment of problems involving rates of change science had to wait for the invention of the methods of the infinitesimal calculus before this disability of the human mind could be circumvented.

But the moving point occupies *successively* a number of different positions in space. If it is a material point that we observe to move from one place to another, we perceive that a certain interval of our duration corresponds with the change of position of the point. Duration was not used up in the occupancy of the different positions O, A, B, C, D, E, and so on, nor in that of the occupancy of the inde-finitely numerous other positions in which we may place the moving point, but in the intervals themselves. We have said "duration" and not "time," using Bergson's term. By duration and time we understand different things.

Time is, for us, only a series of standard events which punctuate, so to speak, our experienced duration. The unit of time is the sidereal day, that is, the interval of time between two successive transits of a fixed star across the arbitrary meridian. But if we try to con-ceptualise this interval we find that we can do so only by breaking it up into smaller intervals, and this we do by using a pendulum of a certain length which makes a certain number of swings (86,400) during the interval between the two transits of the star.

Thus we obtain a smaller interval of duration and we call this a second of time. But for many purposes this interval is too long, and we can again sub-divide it by making use of a tuning-fork which makes, say, 1000 complete vibrations in a second; in this way we obtain still smaller intervals of duration—the sigmata of the physiologists. A sigma, therefore, represents the interval between the beginning and end of one complete vibration of a certain kind of tuning-fork; a second, that between the beginning and end of one complete swing of a pendulum of a certain length, placed at certain parts of the earth's surface; and a day, that between two successive transits of a fixed star across a selected meridian, after all the necessary corrections have been made to the observation. These actual occurrences, the positions of the prongs of the tuning-fork, or those of the bob of the pendulum, or those of the fixed star do not involve duration. We consider the meridian of Greenwich as an imaginary line drawn across the celestial sphere, and the star as a point of light, so that the actual transit is, in the limit, an occurrence which occupies only an "infinitesimal" interval of duration. So also with the pendulum and the tuning-fork; the positions of these things do not "use up" time, and even if the intervals into which we divide astronomical time are indefinitely numerous no real quantity of duration is taken up by their occurrence. We know that the interval between two successive transits of a fixed star are not really constant, that is, the astronomical day is lengthening by an incredibly small part of a second each year, but how do we know this? It is not that we can *feel* the increments of duration, but just that we assume that Newton's laws of motion are true; and hence that the tidal friction due to the motions of the earth, sun, and moon must retard the period of rotation of the earth so that the intervals between two successive transits of a star must become greater.

Thus we do not conceptualise the actual intervals of duration of which we are able to mark the end-points ; they are lived by us, and they are real absolute things independent of our wills. Suppose we come in from a long walk, tired and thirsty, and ask the maid to get tea ready at once. She puts the kettle on the gas stove and then sits down to read. The water takes, say, five minutes to boil. What do we mean by this ?

This is what we mean :—

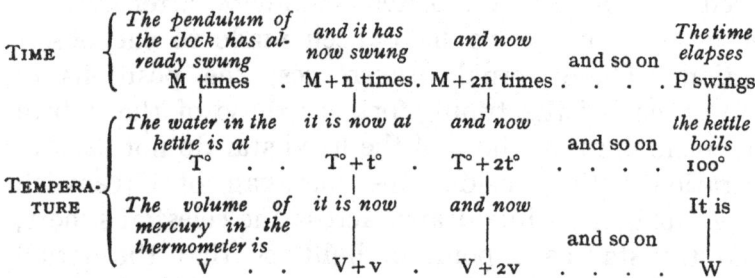

What we call time here is only a series of simultaneously occurring events. The standard events are the positions of the hands of the clock on the clock face, that is, lengths of arc recording the number of swings of the pendulum that have occurred since the beginning of the operation of the boiling of the kettle. When this began, the hands of the clock were at, say, 4.30, and the temperature of the water was then, say, 17° C. ; and, when it ended, the hands of the clock were at 4.35 and the temperature of the water was 100° C. It is only the simultaneities of these events that we have recorded and not the interval of duration that they mark. It does not matter how many times we might have looked at the hands of the clock and the thermometer, we should still have observed only simultaneities.

But we had to *wait* for the kettle to boil, and the temperature 100° was attained *after* the temperature 90°, and so on. What does this mean? While we were waiting, the water seemed to take an intolerably long time to boil. But the maid was reading one of Mr Charles Garvice's novels, and " before she knew where she was " the kettle boiled over. There was a certain interval of duration experienced by her, and another, but different, interval of duration experienced by us. In each case there was a stream of consciousness. We felt fatigue, thirst, a lack of satisfaction, wandering attention, and irritation—all that was our duration. But the maid was identifying herself with Lady Mary, who had sprained an ankle and was being helped along by the new, young gamekeeper, and that was her duration.

There need not be any succession of events in the conceptual representation of a physical process. There is, for instance, no succession in such a conception as is represented by the following diagram—a conception well worth analysis :—

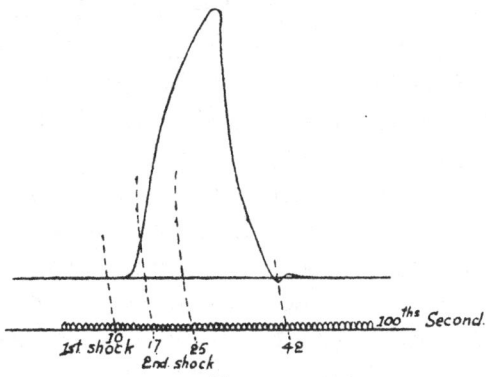

FIG. 7.

The figure represents a tracing made by a muscle-nerve preparation. A living muscle taken from an animal has been attached to a light lever, the end of which makes a scratch on a piece of smoked paper. The paper is fastened on a revolving cylinder and so long as the muscle is motionless the end of the lever marks a horizontal line on the paper. But if the muscle is stimulated so that it contracts and then relaxes again the lever is pulled up and is then lowered, and so its point makes a curve on the paper. The nerve going to the muscle can be stimulated electrically and the moment of the stimulation can be recorded by another lever, which makes a mark on the paper below the trace made by the lever which is attached to the muscle. Two such shocks have been applied to the nerve and they have elicited two contractions of the muscle and these two contractions have fused together.

In the actual experiment the operators could *see* that the muscle moved, and they could *feel* that a certain interval of their own duration coincided with the interval between the first and second depressions of the key that made the electric shocks. But the extent of motion of the muscle was too small, and the depressions of the key succeeded each other too rapidly to be easily observed, and therefore all these events were made to record themselves on the myogram. The series of little notches at the base of the figure represent the movements of the time-lever, that is, they are scratches made on the paper by a little lever which moves up and down at a rate fixed beforehand. Now when this time lever had made ten notches on the paper the first shock was applied to the nerve, and at the eleventh the muscle began to contract. At the seventeenth notch the second shock was applied and the muscle continued to contract. At the twenty-fifth notch the muscle ceased to contract and began to relax, and at the forty-second notch the muscle had ceased to contract. Everything now becomes clear and easy to represent mentally ; the time-lever makes 100 notches on the paper in a second, so that there was an interval of 0.07 seconds between the two stimuli, and these two stimuli produced a compound contraction of the muscle lasting for 0.1 second. This is what the experimenters might have perceived, had human unaided senses been sufficiently acute. But they are not, and so the crude perception of the results of the experiment is replaced by a conception of the train of events involved in the operation. Duration and succession disappear and the myogram represents only a series of simultaneous events of this nature ; the first stimulus occurs simultaneously with the tenth movement of the time-lever ; the second stimulus with the seventeenth, and so on. In seeing the experiment the operators had to *wait* for one phase to be completed before they could observe another one, but in reasoning about it all the phases are spread out and are present in the conception at once. The duration was in the operators but not in the experiment : it was experienced, but it disappears when the results of the experiment are conceptualised.

A succession of events is in ourselves and not in the events observed. If a point is said to move along the locus OD through the positions A, B, C, it is we that have the feeling of succession, and the whole trajectory, or locus, or path of the point corresponds with a portion of our duration. The operation of boiling the kettle corresponds with a portion of our duration, which in its turn corresponds with that part of our duration which was marked by the positions of the hands of the clock. Thus we perceive a simul-

c

taneity in these two trains of events, and this enables us to assign a certain period of astronomical time to the operation of raising the temperature of the water, in the conditions of the experiment, from 17° C. to 100° C. But there is nothing absolute in this interval of astronomical time : what is absolute is that certain successions of events always correspond with other successions of events. A certain number of swings of a seconds-pendulum always corresponds with a certain rise in temperature of a definite mass of water which is in thermal contact with an indefinitely large reservoir of heat at a certain temperature, and, no matter how often we repeat this experience, the same simultaneity is always to be observed. Thus what the physicist considers is not intervals of his own duration but series of correspondences—that is, correspondences of certain standard events with the events which he is studying.

In reality time, in the sense of the astronomer's time, does not enter into the methods of the mathematical physicist. Let us suppose that he is investigating the change that occurs in a material system between the two moments of time t_1 and t_2, these moments being separated from each other by a period of duration that we can feel. Let the system be, say, the earth and moon ; the first body being supposed to be motionless, and the second being supposed to have a certain tangential velocity of movement. If the interval t_1 to t_2 is really an interval of astronomical time, the problem, what is the difference of position of the moon owing to the gravitation of the earth, is incapable of solution, and even if we reduce the interval of time indefinitely while still supposing that it is a finite interval, the mathematical difficulty remains. We then replace the finite interval t_1 to t_2 by the differential dt, which means that the two phases of the system, motionless earth and moving moon at the time t_1, and motionless earth and moving moon at the time t_2, are separated by an interval of time dt, which is smaller than any finite interval that we can conceive. We must then integrate the differential of the position difference so as to obtain the real difference in the condition of the system after the finite interval of time t_1 to t_2 has elapsed. Thus mathematics, incapable of dealing with real intervals of time, *evades* this difficulty by considering tendencies, not real occurrences.

Things that happen in a part of inorganic nature arbitrarily detached from the rest, and investigated by the methods of mathematical physics, do not endure. Let us suppose that we take some silver and add nitric acid to it : the metal dissolves. We can then add hydrochloric acid to the solution and precipitate the metal in the form of chloride ; and we can then fuse this chloride with carbonate of soda, or some other substance, and so obtain the metal again. If we work carefully enough we can repeat this series of operations again and again and the original portion of silver will remain unchanged both in nature and in mass. All the chemical reactions into which it has entered have not affected it in any way ; that is to say, these reactions have not endured.

If we inject a serum, containing a toxin, into the blood stream of a susceptible animal, certain things happen. The animal will become ill, but, provided that the amount of serum which has been injected was not too great, it will recover. If the toxin be again injected a reaction occurs, but the animal does not become so ill as on the first occasion, and after a number of injections the dose administered may be so great as to kill a susceptible animal but may yet produce no effect on the animal which is the subject of the process of immunisation : immunity has been conferred on it. Now can we compare the two operations, that of the solution and precipitation of the metal and that of the immunisation of the animal? We can to some extent, but the analogy soon fails, and indeed we should not attempt to formulate a theory of immunity on a physico-chemical basis if we did not start with the assumption that the series of operations was one in which only physico-chemical reactions were involved, that is to say, there is nothing in the phenomena of immunisation that suggests that what occurs in the animal body is similar to what we can cause to occur in inorganic materials outside the tissues of the living organism. We start with the assumption that the administration of the toxin causes the formation of an antitoxin in very much the same sort of way as the administration of hydrochloric acid to a solution of nitrate of silver causes the formation of chloride of silver. This antitoxin then neutralises the dose of toxin which may be administered after the process of immunisation has been effected, very much in the same sort of way as a certain amount of some acid can be neutralised by an equivalent amount of some base with which the acid can combine. If the reader will analyse any of the theories of immunisation current at the present day he will find that these are the physical ideas that are involved in it.[1] But physiological science has the much more formidable task of explaining the persistence of the immunity. The animal rendered immune to the toxins produced by certain species of bacteria may remain so for many years, that is, for a very long time after the antitoxins originally produced by the reaction of the tissues to the

---

[1] Except that, of course, the reactions that are supposed to occur are very complex ones.

toxins first administered have disappeared. We must imagine, therefore, that the anti-substances produced originally by the reaction of the toxin are produced again and again by the tissues of the susceptible animal, for the latter may resist repeated infections, that is, repeated doses of toxin, without illness. But then the tissues of the animal body are transitory substances and they do not persist unchanged. Muscles, glands, connective tissues, even nerve-fibres and nerve-cells undergo metabolism, and the chemical substances of which they are composed break down into the excretory products, pass out into the blood stream, and are eliminated from the body; while at the same time these tissues are continually being renewed from the nutritive substances in the blood and lymph. It is the *organisation* of the tissues— their form and modes of reaction—that endure, but the material substances of which they are composed are in a state of continual flux. Yet the organisation of these tissues does not persist unchanged, for it is continually *responding* to new conditions experienced by it. The reactions that occur when a toxin is administered to a susceptible animal affect the organisation of its tissues in such a way that the latter acquire the capability of producing antitoxins which may—if we like to say so—neutralise the toxins that enter into them when they become infected. The reaction endures. But this is a different thing from saying that the process is a physico-chemical one alone.

This is what we must understand by the duration of the organism. Everything that it experiences for the first time persists in its organisation. It acquires the ability of responding to some stimulus by a definite, purposeful reaction, the effect of which is to aid it in its struggle for existence; and this reaction, once carried out, becomes a "motor habit" or the basis of a reflex, or in some other way, as in the process of immunisation, remains a part of the modes of functioning of the animal. In our behaviour certain cerebral nerve tracts become laid down and continue to exist throughout life, modifying all our future experience. Our past experience accumulates. There must be direct continuity in our flux of consciousness, for no perception seems ever to fade absolutely from memory. This continual addition of perceptions to those that already exist makes our consciousness ever become more complex, so that a perception experienced for the first time is never quite the same when it is again experienced. The first time that we go up and down in an elevator, or sit on a " joy-wheel," or ascend in a balloon or an aeroplane, or become intoxicated, constitutes an unique event in our lives, and we experience a " new sensation." What the blasé man of the world complains of is this accumulation, or rather persistence, of his experiences. A repetition of the same stimulus never again begets the same perception. The first hearing of a modern drawing-room song may be enjoyable, but the next time we hear it we are not interested, and by-and-bye it becomes very tiresome. The first hearing of a great symphony usually perplexes us, and we are perhaps repelled by unusual harmonies, or progressions, or strange modulations, but subsequent hearings afford increasing pleasure. We say that there was " so much in it " that we did not understand it, yet precisely the same series of external stimuli affected our auditory membranes on each occasion, and the same molecular disturbances were transmitted along our afferent nerves to the central nervous system, where the same physical effects must have been produced. The difference in all these cases between the repetitions of the same stimuli was that the later ones became added to the earlier ones, so that the state of consciousness produced by, or which was concomitant with, these external stimuli was a different state in each case.

This is the duration of the intelligently acting animal : it is not merely memory,.but memory and the accumulation of all its past modes of responding to changes in its environment, whether these modes of response were conscious ones (as in the case of an intelligently performed or " learned " action), or unconscious ones (as, for instance, in the case of the acquirement of immunity by an animal which had become able to resist disease). It is not merely the experience of the individual organism, but also all the experience of those things which were done or experienced by the ancestry of the organism, and which were transmitted by heredity to the progeny. Motor habits are formed, so that much the same series of muscular actions are carried out when a stimulus formerly experienced is again experienced. Pure memory remains, so that the images of past things and actions somehow persist in our consciousness. Physical analogy suggests that these images are inscribed on the substance of the brain or are stored away in some manner ; but, apart from the incredible difficulty of imagining a mechanism competent for this purpose, it is obvious that we thus apply to the investigation of our consciousness (which is an intensive multiplicity), the concept of extension which can only apply in all its strictness to the things outside ourselves on which we are able to act. All these motor habits, functional reactions, and memory images are our duration or accumulated experience. The motor habits and those functional habitual reactions of other parts of the body than the sensori-motor system are the basis of our actions, but the memory images are, so to speak, pressed back into that part of our organisation which does not emerge into consciousness. Only so much of them as bear on the situation in which we, for the moment, find ourselves and which may therefore influence our actions, flash out into consciousness. As " dreamers " we indulge ourselves in the luxury of becoming conscious of these memory images, but as " men of action " we sternly repress them, or so much of them as do not assist us in the actions that we are performing. Yet it is in the experience of each of us that, in spite of this continual inhibition, parts of our memories slip through the barriers of utility and surreptitiously remind us of all that we have been and thought.

Thus we simplify the stream of our consciousness. That of which we are conscious at any time is never more than a part of our crude sensation : we never perceive more than a small part of all that our organs of sense transmit to our central nervous system. But even these chosen perceptions of the external world are so rich, so chaotic and confused, that we are unable to attend to them all at once and we therefore " skeletonise " the contents of our consciousness. We think about it a bit at a time. It is an unitary thing, unable to be broken up, but we look at it from a great number of different points of view, so to speak ; and then, fixing our attention on some aspect of it, we agree to ignore all the rest. We thus detach parts of it from the rest and, having thus arbitrarily decomposed it, we call these separate aspects the elements of our perceptions, and confer upon them separate existence in space and time. We remember and classify things and group together all those that seem to resemble each other. We form genera, agreeing to ignore all but the most general characteristics of the things which we try to conceptualise. We do not think separately about all the dogs or horses or fishes that we have ever seen, but we group all these animals into species, and it is usually the species that we think about when the idea of a dog or a horse or a herring emerges into our consciousness. When we think about a tramcar we do not think about all the separate vehicles that we have seen, nor about their colours, nor the advertisements on the boards outside, nor the people hanging on to the straps inside. Just so much of the experience of what is relevant to the purpose of our thought enters into our idea of the tramcar : it is a conceptual vehicle that we think about. Such is the nature of the concepts that form the basis of our reasoning : they are generalised aspects of our experience of nature, usually poorer in content than were the actually perceived things, except when it is necessary that some individual thing seen or otherwise experienced should be investigated or reasoned about. All our descriptions of nature are conceptual schemes. The world of perception, says William James, is too rich to be attended to all at once, but in conceptualising it we spread it out and make it thinner, and we mark out boundaries and division lines in it that do not really exist. It is this generalised nature that is the subject matter of our reasoning of pure science ; and it is these concepts that form the matter of all our descriptions. We do not describe nature " as we see it," it is our conceptions that we write about. Genera and species and varieties do not really exist in the animate world : all these are logical categories generated by our thought, concepts that facilitate our descriptions. When an anatomist gives an account of the structure of an animal he does not say what it looks like, nor as a rule does he content himself by making a photograph of his dissections. For him the animal is a complex of muscles, skeleton, nerves, glands, and so on, and in his drawings all these things are given an individuality that they do not really possess. In the living creature there were no such sharply-distinguished organs as a good drawing represents : all are bound together and are continuous. But for practical convenience in description—that is, in the long run, that we may *act upon* these things, we isolate from each other aspects that are in reality one unitary whole.

The universe, that is, all that is given to us, presents itself as immediately perceived phenomena which are then conceptually transformed. It is an aggregate of things—gross matter, particles, molecules, atoms, and electrons. These things have separate existence and shape, so that each of them lies outside all other things —we apply to them the category of extension. They possess properties—that is, they are hard, or heavy, or hot, or cold, or they are coloured, or they smell, and so on—we thus apply to them the category of inherence. They are not things that are immutable, for they change in place, or are transformed in other ways, that is, they are acted upon by energies. But beneath the properties of the things, or the transformations that they undergo, we imagine something that has properties and which transforms : it is not convenient that we speak solely of attributes or transformations as entities in themselves, for we think of things as having properties and being subject to transformations. Thus we apply the category of substance.

Has this universe that we construct from the data of sensation objective reality ? We are led quite naturally by our study of physiology to the notion of idealism. We see that our perception of things, that is, our knowledge of the universe, depends on the integrity of functioning of certain bodily structures, and upon the condition that in men in general this integrity of functioning is normal, that is, common to the great majority of mankind.

To say that a thing exists is to say that it is perceived in some way ; that immediately or remotely it affects our state of consciousness. To say that the star Sirius exists is to say that the stimulation of the retina by a minute spot of light transmits certain molecular disturbances along the optic nerve, and that other molecular disturbances are set up in the tissues of the central nervous system. Even if we do not see those dark stars that we know to exist, there are still evidences of their being that in some way affect the instruments of the astronomer and lead to their being perceived. Even if we do not actually see the emanations from a radio-active substance, we can cause these emanations to produce changes in something that we can see. We speak of the star as a minute spot of coloured light. But if we are short-sighted the spot becomes a little flare, and if we are colour-blind the hue of the star is different from what it is to normal persons. If we put a drop of atropine into one eye and then close the other, objects appear to lose their distinctness, but if we close this eye and then open the other, the original sharpness of vision returns. When we are bilious, wisps and spots may appear on a sheet of white paper that at other times was blank. If we take an overdose of quinine, rustlings and singing noises become apparent even in conditions that ought to preclude all sensation of sound. If we have a bad cold, we do not smell substances which at other times strongly affect our olfactory membranes. When we become intoxicated, a host of aberrations of sense displace our normal perceptions of things.

Our perception of the universe, therefore, depends on the normal functioning of our organs of sense, that is, such modes of functioning as we can describe and communicate to others, and which are thus common to the majority of other men and women. These perceptions resulting from the normal functioning of the organs of sense constitute givenness, and we enlarge, or conceptualise this givenness and call it the subject matter of science. But what is this reality that we say is external to us ? It is, we see, our inner consciousness. If we walk along a road in the dark we can feel what is the nature of the path on which we tread, whether stones or gravel, or sand or grass. But this feeling is obviously not in the soles of our boots, and neither is it in the skin of the feet, for we should feel nothing if the afferent nerves in the legs were severed. Is it then in the brain ? It would appear to be there, but it disappears if certain tracts in the brain are injured.

All that we can say is that the appearance of reality of things outside ourselves is only the ever-changing condition of our consciousness. This is all that we immediately know, and if we say that there is an universe external to ourselves we thus project outside our own minds what is in them ; and we construct an environment which may or may not exist, but which we have no right to say does exist. A philosophy based on the science of the organism would appear to be restricted to this idealistic view of the universe. When we come across it for the first time when we are young it appeals to us with all the force of exact reasoning, and yet it has all the charm of paradox. There is no part of our intuitive knowledge which appears to us to be more certain than this distinction between ourselves and an outer environment : we *know* that our conscious Ego is something different from our body— and we know that outside our body there is something else. Yet the idealistic view so appeals to the intellect that we cannot think speculatively about it without, at times, almost convincing ourselves of the unreality and shadowiness of all that at other times seems most real and tangible ; and we indulge in these speculations all the more readily because we know that whenever we begin to *act,* the intuitively felt body and outer world will return to us with all their original conviction of reality.

Some such system of idealism must generally characterise a system of philosophy founded on pure reasoning. We cannot but feel that the universe that we construct is one that depends on our perceptions : it is our perceptions. The essence of a thing is that it is perceived. If there were no mind to perceive it, would it exist ? The universe is our thought, and we, that is our thought, exist only in the Thought of an absolute Mind which we call God. Such is the metaphysics to which the study of sensation led Berkeley.

The metaphysics of science has taken another turn. It is true that men and women see something outside themselves which differs slightly in different individuals —these differences are due to what we call the " personal equation." The image of the universe seen by some individuals may differ profoundly from the image seen by some others, or most others ; but a well-marked gap separates these slight individual deviations in the images seen by normal individuals from the large deviations seen by those whose perceptions are what we call pathological ones. The normal universe common to the majority of men and women is an aggregate of molecules in motion. But this is a conclusion with which modern physics has been unable to remain content, for molecules must be able to act on each other across empty space, and this is inconceivable. The universe therefore consists of a homogeneous immaterial medium, the ether of space, and this is the true *substantia physica*. Molecules and radiation are conditions of the ether, and for the physicist it is the only reality. The " materialism " of our own time is therefore the belief in the existence, unconditioned by time or anything else, of the ether, or physical continuum ; a homogeneous medium, of which matter and energy, and the consciousness of the organism, are only states or conditions.

The materialism of the twentieth century, like the idealism of Berkeley, thus finds that there is something outside our own consciousness that possesses absolute existence. To the materialist it is the ether of space, and to Berkeley it is the existence of absolute Mind. But if our desire to avoid metaphysics is a genuine one, we must reject the notion of the universal ether no less than we must reject the notion of an absolute Mind, and we must rest content with pure phenomenalism. For each of us there can be no existence except that which is perceived or conceptualised. There is nothing but our own consciousness ; there cannot even be an Ego which perceives ; there is only perception. We never do really believe this in spite of our professions of reason. We find on strict self-analysis that we believe that there is an Ego that perceives and that there are other Egos that perceive, and that the universe which our Ego perceives is also the same universe that other Egos perceive. If we did not believe that there were other men and women that perceived—other consciousnesses like our own, all that part of our own behaviour that we call morality would be meaningless. In a philosophy of pure idealism other men and women are only phenomena ; only bodies moving in nature. Why, then, should these elements of our consciousness influence the rest of our consciousness as if they were men and women like ourselves. All this amounts to saying that while our speculative thought suggests to us that all that exists is our stream of consciousness, our actions must convince us that there are other thinking individuals like ourselves.[1]

Even if we do surrender ourselves to phenomenalism and try to believe that all that exists is our own consciousness, the fact of our duration would suggest to us that this present consciousness is not all. Our reality is not only that which is present in our minds now, but all that was ever present in our mind. All that we have ever thought and done persists and forms our conscious and unconscious experience. This past of ours is something that is ever being added to, or becoming incorporated with, our present state of consciousness ; and if it is something other than that which we now perceive and conceptualise, it is something that has an existence of its own.

We must believe that there is something that we perceive, and not that we merely perceive. For the phases of our immediate givenness, that is, those things which were present in our minds from moment to

---

[1] The reader may recognise in this argument that of Driesch's *Three Windows into the Absolute*.

moment of the past were connected together and had direction, and this direction was something that could not be influenced by our will, and may even have been contrary to our will. Something that is very hot always cools, a wheel that is revolving of itself always comes to a stop, a pendulum ceases to swing, a stone that is rolling down a hill continues to roll. Let us look back at a fire that was going out : it is now nearly dead ; let us start a pendulum to swing and then go away : when we come back the pendulum is still swinging but the amplitude of its vibrations is now less than it was ; let us look away from the stone that was falling : when we look again it is still falling but it is not where it was. In all our givenness, in all the phenomena that we perceive, there is something that is determined and unequivocal, something that goes its own way apart from our consciousness of it.

Above all, we have the conviction of absoluteness in our sense of personal identity. We, that is our Ego, are something that endures, and we can trace no beginning to our identity, and we have no intuition that it will cease to exist. Our Ego is now the same Ego that it was in the past, and round it something has accumulated—the memories of our former perceptions, and the habits that these have engendered. Did our Ego create this from itself ? Was it not rather a centre of action which, residing in an existence other than itself—the absolute which we call the universe— modified that existence and continually acquired new relationships to it ?

CHAPTER II

THE ORGANISM AS A MECHANISM

WE propose now to consider the organism purely as a physico-chemical mechanism, but before doing so it may be useful to summarise the results of the discussions of the last chapter. Let us, for the moment, cease to regard the organism as a structure—a " constellation of parts "—and think of it as the physiologist does : it is a machine ; it is essentially " something happening." What, then, is the object of its activity ? Whatever else the study of natural history shows us, it shows us this, that the immediate object of the activity of the organism is to adapt itself to its surroundings. It must master its environment, and subdue, or at least avoid whatever in the latter is inimical. It must avoid accident, disease, and death, it must find food and shelter ; it must seek for those conditions of the environment which are most favourable to its prolonged existence. Ultimate aims—the preservation of its race, ethical ideals—do not concern us in the meantime. The main object of the functioning of the individual organism is that it may dominate its environment, and obtain mastery over inert matter. Consciously or unconsciously it acts towards this end.

All those actions which we call reflex, or automatic, or instinctive, have this in common, that the organism in performing them comes into relation with only a very limited region of its environment. But knowing that

D

region intuitively, its actions have a completeness that an intelligent action does not exhibit until it has become so habitual as to approach to automatic acting. The relations between the organism and that part of its world on which it acts, intuitively or instinctively, is something like that between a key and the lock to which it is fitted : it opens this lock, perhaps one or two others which resemble it, but no more. Now just because of this perfect, but restricted, adjustment of the instinctive or automatically acting organism to the objects on which it operates, knowledge of all else in the environment becomes of little consequence.

It is clear that intelligent acting involves deliberation. The almost inevitable motor response to a stimulus, which is characteristic of the reflex or instinct, does not occur in the intelligent action : instead of this we find that we choose between two or more responses to the same stimulus. We reply to the latter by doing *this* now, and *that* another time ; and we see at once what results flow from acting differently upon the same part of our environment, or acting in the same way upon different parts. Perception, that is, knowledge of the world, arises from acting ; and as our actions, when carried out intelligently, become almost infinitely varied, the environment appears to us in very many aspects. In every action we modify that part of our surroundings on which we operate. We can produce many modifications that are of no use to us : these we do not attend to. We produce others that are useful, and then we note the sequences of events involved in our actions. Thus we discover or invent natural law—an environment which is an orderly one. We can calculate and predict what will happen : we produce, for instance, a *Nautical Almanac,* at once the type of useful knowledge and of knowledge of sequences of events rigidly determined—knowledge in short that is mechanistic; and which has been engendered by the necessity for acting on our environment in our own interests.

All this, the reader may note, is Bergson's theory of intellectual knowledge, a theory which, new and paradoxical at first, becomes more and more convincing the longer we think about it, until at last it seems so obvious that we wonder that it ever seemed new. Our modes of thinking become constrained into certain grooves, just because these modes of thinking have been those that were generated by our modes of acting. So long as our thinking relates only to our acting, its exercise is legitimate. But if its object is pure speculation its results may be illusory, for a method has been applied to objects other than those for which it was evolved. Let us now extend our intellectual methods to the investigation of the organism. Necessarily we must reason about the latter as a mechanism if we reason about it at all.

If it is a mechanism it must conform to the laws of energetics, for science, so far as it is quantitative, whether its results are expressed in the form of equations or inequalities, is based on these principles.

The first principle of energetics,[1] or the first law of thermodynamics, is that of the conservation of energy. Let us think of an isolated system of parts such as the sun with its assemblage of planets, satellites, and other bodies : in reality these do not form an isolated system, but we can regard them as such by supposing that just as much energy is received by them from the rest of the universe as is radiated off by them to the rest of the universe. In this system, then, the sum of a

[1] See appendix, p. 356.

certain entity remains constant, and no conceivable process can diminish or increase its quantity. We call this entity energy, and we usually extend the principle of its absolute conservation to matter, though this extension is unnecessary, for we must think of matter in terms of energy. Stated more generally the principle is that whatever exists must continue to exist, if we are to regard this existence as a real one.[1]

It is not at all self-evident to the mind that energy must be conserved, for we see that, to all appearance, it may disappear. A golf-ball driven up the side of a hill possesses energy while in flight, kinetic energy or the energy of motion ; but this apparently is lost when the ball alights on the hill-top and comes to rest. We say, however, that it now possesses potential energy in virtue of its position ; for if the hill is a steep one a little push will start the ball rolling down with increasing velocity, and when it reaches the spot from which it was originally impelled it possesses kinetic energy. This is described as one-half of the mass of the ball multiplied by the square of its velocity. Now the kinetic energy of the ball at the instant when it left the head of the driver ought to be equal to its kinetic energy when it reached the same horizontal level on its downward roll. Yet it can easily be shown that this is not the case, and we account for the lost kinetic energy by saying that it has been dissipated by the friction of the ball against the atmosphere in its flight, and against the side of the hill on its roll back. We cannot verify this quantitatively, but we are quite certain that it is the case. If we take a clock-spring and wind it up, the energy expended becomes potential in the spring, and when the latter

---

[1] The principal reason why we do not believe in phantasms is that these appearances *are not conserved.*

is released most of it is recovered. But we may dissolve the spring in weak acid without allowing it to uncoil. What then becomes of the energy imparted to it? We are compelled to say that it has changed the physical condition of the solution into which it passes, either becoming potential in this solution, or becoming dissipated in some way. Yet again we cannot trace this transformation experimentally though we may be quite sure that all the energy potential in the coiled spring is conceivably traceable. Suppose, again, we burn some hundredweights of coal in a steam-boiler furnace. Heat is evolved which raises steam in the boiler, and the steam actuates an engine, and the latter exhibits measurable kinetic energy. Where did this come from? It was potential in the coal, we say, though no method known to physics enables us to prove this by mere inspection of the coal. We must cause the latter to undergo some transformation. But by rigid methods we can estimate very exactly the potential energy of the coal, and we can calculate the kinetic energy equivalent to this. Yet again we find that the kinetic energy of the steam-engine is only a fraction of that which calculation shows us is the equivalent of the kinetic energy of the coal. What becomes of the balance? We can be quite certain that it has been dissipated in friction, radiation, loss of heat by conduction, loss of heat in the condenser, and so on, although we cannot prove this rigidly by experimental methods.

Think of the universe as an isolated system. It contains an invariable quantity of energy. This energy may be that of bodies in motion—suns, planets, cosmic dust, molecules, etc.—when it is kinetic energy ; or it may be the energy of electric charges at rest or in motion ; or any one of the many kinds of potential energy. It may pass through numerous transformations—the chemical potential energy of coal may be transformed into the kinetic energy of water molecules (steam at high temperature), and this into the kinetic energy of the revolving armature of a dynamo, and this again into the energy of moving electrons (the current of electricity in the circuit of the dynamo), and then again into the energy of ethereal vibration (light, heat, X-rays, or other electro-magnetic waves), and these again into mechanical or kinetic energy, and so on.   When we say that we can control energy we say that we can produce these transformations ; we can *cause* things to happen, we bring *becoming* into being.   In this sense energy is causality. But while the sum-total of energy in the universe remains constant, the sum of causality continually diminishes.   Energy is the power, or condition, of producing *diversity*, but while energy can suffer no diminution of quantity, diversity tends continually to decrease.

In the last two sentences we state, in one way, the second law of thermodynamics—in some respects the most fundamental result of our experience in the physical investigation of the universe.   In its most technical form, as enunciated by Clausius, this law states that the value of a certain mathematical function, called *entropy*,[1] tends continually towards a maximum, when it is applied to the universe as a whole.   When we say the "universe," we mean all that comes within our power of physical investigation.   Let us now see what this statement means.

---

[1] See appendix, p. 369.   Entropy is a shadowy kind of concept, difficult to grasp.   But again we may point out that the reader who would extend the notion of mechanism into life simply *must* grasp it.

The energy of the solar system is in part the kinetic energy of those parts of it which are in motion—planets, planetesimals,[1] and satellites. This quantity of energy is enormously great. In the case of our earth it is $\frac{1}{2}mv^2$, m being the mass of the earth, and v its velocity. Translated into numerical symbols we find this quantity almost inconceivable. The greater part of this energy is *unavailable*, that is, it can undergo no transformations. But because the earth is in rotation at the same time as it revolves round the sun, and because the moon revolves round the earth, there are tides in the watery and atmospheric envelopes of the earth. The energy of the tides is the kinetic energy of water or air in motion, and we can employ this energy in the production of transformations, and it is therefore *available*. But well-known investigations have shown that the tides produce friction, and that the period of rotation of the earth is slowly becoming greater. Ultimately the earth will rotate on its own axis in the same time that it revolves round the sun—then a year and day will be of the same length. When that occurs, the sun, earth, and moon will be in equilibrium, and tidal phenomena due to the sun will cease. The kinetic energy of the earth, rotating once in 24 hours is obviously greater than its kinetic energy when rotating in the period which will then be its year. What has become of the balance? It has been transformed into the mechanical friction of the tides against the surface of the earth,[2] and this friction has been transformed into low-temperature heat, and this heat has been radiated off into space.

[1] Meteorites, cosmic dust, and other small particles moving in the solar system within influence of the sun's gravity.

[2] Not entirely, of course, but whatever be the transformation it ends in heat production.

The solar system also contains energy in the form of the heated sun and planets, and in the form of chemical potential energy of the substances of which those bodies are composed. Let us think of the system, sun and earth. The sun contains enormous heat energy, its temperature being some 6000° C. absolute.[1] It contains enormous chemical energy in the shape of compounds existing beneath its outer envelopes, and it contains energy in the form of its own gravity—its contraction together produces heat. But this heat is being continually radiated away: chemical reactions must occur in which the potential chemical energy of its substances must become transformed into heat, and this heat is also radiated away; contraction of its mass must occur up to a point when the materials are as closely packed together as possible; heat is developed during the contraction, and this also passes away by radiation. Suppose that modern speculations are well founded and that radio-active substances are present in the sun: in the atomic disintegration of these substances heat is produced and again radiated. Therefore in whatever form energy exists in the sun, it transforms into heat and this radiates. The ultimate fate of the sun is to cool down and solidify. It will then move through space as a body having a cool, solid crust, and an intensely heated interior. Slowly, very slowly, this heated interior will cool down by the conduction of its heat from the core to the outer shell, and by the radiation of this heat from the shell into space. For incredibly long periods radio-active substances in the interior must generate heat, but even this process must reach an end.

The energy received by the earth is that of solar

[1] Absolute temperature is Centigrade temperature + 273. This is, of course not a full definition, but it is sufficient for our present discussion.

and stellar radiation. Stellar radiation is minute, the absolute temperature of cosmic space (or ether) being about $-263°$ C. The absolute temperature of the earth is about $+17°$ C., so that it radiates off more heat into space (other than that represented by the sun) than it receives. All energy-transformations on the earth (except tidal effects, and energy-conduction from the heated core, and possibly radio-active effects) are transformations of this solar energy received by radiation. We see these in oceanic and atmospheric circulations (currents, winds, rainfall, etc.). We see them also in the transformations of the chemical potential energy of coal and other products of life— products in which the contained potential energy has been absorbed from solar radiation.

Let us follow the transformations of this energy. Oceanic currents transport heat from the equatorial sea-areas to the colder temperate and polar areas, and compensatory polar currents flow towards the equator, absorbing heat from the waters of temperate and equatorial areas. Winds act in an analogous way. Water is evaporated where the solar radiation is intense, and heat is absorbed in the transformation of water into aqueous vapour. Then this water vapour is transported in the winds into regions where it becomes condensed and precipitated as rain or snow, heat being emitted in this condensation. In all these movements there is friction, and this friction transforms to heat. In all the effect is the general distribution over the earth of the heat which the equatorial regions receive in excess of that which the polar regions receive. Other mechanical effects are also produced by oceanic and atmospheric circulations—the denudation of the coasts by tides and storms, the erosion of the land by rivers, rains, snow, and ice, the transport of dust in winds, etc.

In all these friction is produced, and this friction passes into heat.

The potential chemical energy which results from absorption of solar radiation by plants is principally accumulated as coal. Apart from the interference of man, this coal would slowly accumulate, perhaps it would more slowly disappear by bacterial action, or by physical transformations. In these transformations the energy of the coal would become heat energy and the potential energy of the gas produced by bacterial activity. By man's agency the coal suffers other transformations, and in the present phase of civilisation it is his chief source of energy. It is available for doing work of many kinds, and in all these forms of work it becomes transformed by chemical action (burning) into high temperature heat.

We can cause this potential energy of coal to transform into mechanical energy of machines, vehicles, and ships in motion by causing it to pass into heat. In the steam-engine, or gas-engine, a highly heated gas (steam, or the mixture resulting from the explosion of coal gas and air in the cylinder of the engine) expands and propels a piston or rotates a turbine. (Obviously in the petrol engine the same essential process takes place.) We employ this kinetic energy directly in transport, or we cause it to undergo other transformations. In the dynamo, kinetic energy of machinery in motion transforms to electrical energy ; and this may transform to radiant energy (light, heat in electric radiators, wireless telegraphy radiations), or it may transform to chemical energy (the manufacture of carborundum in the electric furnace, for instance), or it may transform again to the kinetic energy of bodies in motion (electric traction). In innumerable ways the human power of direction causes transformation of this accumulated potential energy, and the reader will notice the analogy of all this with the essential, unconsciously expressed activity of the animal organism in its own metabolism—a point to which we return later.

Notice now that all the energy-transformations we have noticed are *irreversible*. This is a matter of deep philosophical importance, and we must devote some time to it. Consider first of all the working of the steam-engine ; what occurs is this—coal is burned in the boiler-furnace, that is to say, potential chemical energy passes into heat and this vaporises water in the boiler, producing a gas at high temperature (steam). This gas expands in the high-pressure cylinder of the engine, driving forward a piston; it expands further in the intermediate cylinder, propelling its piston also, and again in the low-pressure cylinder. It is then cooled by passing through the condenser, and in the contraction further mechanical energy is obtained. The train of events thus begins with a gas at a high temperature and ends with the same gas at the temperature of the water in the condenser. The heat lost is transformed into the mechanical energy of the engine. But not all of it. A certain quantity is lost by radiation from the boiler walls, the walls of the steam-pipes, the cylinders, and other parts of the engine ; also some of the energy is transformed to friction, and this again to heat. In this way a very considerable part of the energy contained in the coal is frittered away in unavoidable heat-conduction and radiation, and a last residue of it goes down the drain, so to speak, with the condenser water. This loss is inherent in the nature of the mechanism of the engine.

Suppose that the energy of the engine is employed to drive a dynamo. The armature of the latter rotates against the constraint of powerful electro-magnets, and in so doing a current of electricity is generated. By the law of conservation this current should contain as much energy as was put into the rotation of the armature ; as a matter of fact it does not, and the deficiency is represented by the friction of the parts of the machine against each other, by imperfect conductivity of electricity in the wires, and by imperfect insulation of the current. Friction, imperfect conductivity, and imperfect insulation all transform to heat, and this radiates away. Suppose now that the current is used for lighting purposes : to do this it must heat the metallic filaments in the lamps, or the points of the carbons in an arc. This heat then transforms to light, but along with the light, which was the object of the transformation, heat is produced, and this heat radiates away.

The actual process in which the particular form of energy required is generated may or may not be reversible in theory. That employed in the steam-engine is not, for if we start with a cold boiler and then work the engine backwards we could not raise steam. The process in the dynamo is theoretically reversible : if we send a current of electricity into a dynamo the machine will begin to rotate, and become a motor, so that we can obtain mechanical work from it. Now in theory all forms of energy are mutually convertible, and all can be expressed in terms of a common unit. The unit of mechanical energy is called the *erg* : let a current, the energy of which is equal to N ergs, be sent into the dynamo, then we ought to obtain from the latter mechanical energy equal to N ergs. Conversely, if N ergs of mechanical energy be employed to rotate the dynamo, we should obtain electrical energy equal to this amount. Now as a matter of fact we do not obtain these theoretical conversions, for some of the electrical energy is dissipated when we employ the machine as a motor, and some of the mechanical energy is likewise dissipated when we employ it as a dynamo.

The entity that we call energy is the product of two factors, a capacity-factor and an intensity-factor. Thus :—

Mechanical energy of water power = quantity of water × height at which it is situated above the water-motor.

Energy of an electric current   = quantity of electricity × electrical potential.

Chemical energy   = equivalent weight of the substance × chemical potential.

What is it that determines whether or not an energy-transformation will occur ?   It is the condition that a difference of the intensity-factors of the energy in different parts of a system exists.   Water will flow from a higher to a lower level, doing work as it flows, if it is directed through a motor.   Electricity will flow if there is a difference of electrical potential.   A chemical reaction will occur if two substances before interacting possess greater. chemical potential than do the products which may possibly be formed during the interaction.   Coal and oxygen possess greater chemical potential than do carbon dioxide and water, therefore they will combine, forming carbon dioxide and water.   Energy-transformations will therefore occur wherever it is possible that differences of intensity or potential can become abolished.   The energy that may thus flow from a condition of high to a condition of low potential, undergoing a transformation as it flows, is the available energy of the system of bodies in which it is contained.   A closed vessel surrounded by an envelope impervious to heat, and containing a mixture of oxygen and hydrogen, is an isolated system containing available energy. Let the mixture be fired by an electric spark, and heat is evolved. The total energy of the system is unaltered in amount, but the available energy has disappeared, since the heated water vapour is incapable of undergoing further transformations while it forms part of its isolated system.[1]

All physical processes are therefore *irreversible*, that is to say, proceed in one direction only. Either a process is irreversible in the sense that it cannot proceed both in the positive and negative directions (a steam-engine, for instance), or it is irreversible in the sense that while it proceeds the energy involved in it becomes less capable of being transformed into other conditions. (In the theoretically reversible dynamo, energy becomes dissipated in the form of heat.) The following statements may be regarded as axioms [2] :—

(1) " If a system can undergo an irreversible change, it will do so."

(2) " A perfectly reversible change cannot take place by itself."

In the phenomena studied by physics we see only

---

[1] It is really necessary to lay stress on the distinction between available and unavailable energy, as it is one which many biologists appear to ignore. Thus, a popular book on the making of the earth attempts to argue that essential distinctions between living and inorganic matter are non-existent. One of these distinctions is that organisms absorb energy, and this author points to the absorption of " latent heat " by melting ice as an example of the absorption of energy in a purely physical process. Consider a system consisting of a block of ice and a small steam boiler. We can obtain work from this by the melting of the ice—that is, its " absorption of latent heat." The system, ice at o° C. + steam at 100° C., possesses available energy, but the system, melted ice + condensed steam, both at the same temperature, contains none. The molecules of water at o° C. " absorb energy," that is to say, their kinetic energy becomes greater, but their available energy in the system has disappeared. In saying that the organism absorbs energy, we mean, of course, that it accumulates available energy, that is, the power of producing physical transformations. (See further, appendix, p. 366.)

[2] Bryan, *Thermodynamics*: Teubner, Leipzig, 1907, p. 40.

irreversible changes. In all these processes energy descends the incline, and some (considerable) fraction of the amount involved passes in*:o conditions in which it is incapable of further transformation ; in all, energy becomes less and less available. Expressed in its most technical form, the second law of thermodynamics states that entropy tends continually to increase. Every such process as we can study in physics " leaves an indelible imprint somewhere or other on the progress of events in the universe considered as a whole." [1]

We cannot observe a truly isolated system. The earth itself is part of the solar system, and the latter receives energy from, and radiates it to the rest of, the universe. Our only isolated system is the whole universe. We must think of it, in so far as we regard it as physical, as a finite system : if it is infinite, our speculations become meaningless. The universe therefore is a system in which energy tends continually towards degradation. In every process that occurs in it—that is to say every purely physical process— heat is evolved, and this heat is distributed by conduction and radiation, and tends to become universally diffused throughout all its parts. When this ultimate, uniform distribution of energy will have been attained, all physical phenomena will have ceased. It is useless to argue that universal phenomena are cyclical. We vainly invoke the speculations (founded on rather prematurely developed cosmical physics) of stellar collisions, light-radiation pressure, the distribution of cosmic dust, etc. to support our notions of alternate phases of dissipation and concentration of energy ; close analysis will show that all these processes must be irreversible. The picture physics exhibits to us is that of the universe as a clock running down ; of an

[1] Bryan, *Thermodynamics*, p. 195.

ultimate extinction of all becoming; an universal physical death.

In this conclusion there is nothing that is speculative. It is the least metaphysical of the great generalisations of science. It represents simply our experience of the direction in which physical changes are proceeding. Based upon the most exact methods of science known to us, nothing seems more certain and more capable of rigorous mathematical investigation.

And yet we are certain that it is not universally true. For there must always *have been* an universe—at least our intellect is incapable of conceiving beginning. If we suppose a beginning, an unconditioned creation, at once we leap from science into the rankest of metaphysics. Holding, then, that the duration of our physical universe is an infinite one, we see that the ultimate attainment of energy—dissipation—must have occurred if our physics is true. It does not matter what new sources of energy modern investigation has shown to us; nor do the incredibly great lapses of duration necessary for the depletion of these sources matter. We have eternity to draw upon. Everywhere in the universe we see diversity and becoming. Is then the whole problem a transcendental one, or is the second law untrue? We refuse to regard the problem as insoluble, and we must think of the second law as true of our physical experience only. But our conception of the universe shows that it cannot be true, and so we have to seek for an influence compensatory to it.

If the organism is a mechanism of the physicochemical kind, it should therefore conform to the two great principles of energetics established by the physicists. Now there can be no doubt that the law of energy-conservation does apply to all the processes observed in animals and plants. Let us consider the " calorimetric experiments." An animal, together with the food and oxygen supplied to it, and the various substances excreted by it, constitutes a physical system. This system can be approximately isolated so that no heat enters it from outside, while the heat that leaves it can be determined quantitatively. The animal is made to perform mechanical work, and this is measured. The energy-value of the food ingested by it, and that of the excreta, can be estimated. All the physical conditions can thus be controlled, and the results of such experiments show that energy is conserved. The energy contained in the food is greatly in excess of the energy contained in the excreta, but the deficit is quantitatively represented by the work done by the animal, and by the heat lost in conduction and radiation from its body. The difference between the observed results and the theoretical ones are within the limits of error of the experiment. The metabolism of the animal as a whole, then, conforms to the law of conservation, and the general results of physiology all go to show that this is also true of chemico-physical changes considered in detail.

It cannot be shown that the second law, that of the dissipation of energy, applies to the organism with all the strictness in which it applies to purely physical systems. If we consider only the warm-blooded animal we do indeed find that its general metabolism does proceed in one direction, and that irreversible changes occur. In the mammal and bird we have organisms which present a superficial resemblance to the heat-engine, with respect to their chemico-physical processes, a resemblance, however, which is rather an analogy than an identity of processes. In the heat-

E

engine we have (1) a mechanism of parts which do not change in material and relationships to each other (boiler, cylinder, pistons, cranks, slide-valves, etc.) ; and (2) a working substance (the steam).

Energy in the form of the chemical potential of coal and oxygen is supplied to the mechanism. The coal is oxidised, producing heat. The heat then expands the working substance (the water in the boiler), and this working substance—now a gas at high temperature and pressure—propels the piston and confers kinetic energy on the engine. Note the essential steps in this process : substances of high chemical potential (coal and oxygen) suffer transformation into substances of low chemical potential (carbon dioxide and water), and the difference of energy appears as high-temperature heat (increased kinetic energy of water molecules, to be more precise). This heat is then transformed into mechanical work (the kinetic energy of the molecules of steam is imparted to the piston of the engine). But in this transformation only a relatively small proportion (10% to 20%) of the energy available is transformed into mechanical work : the rest is dissipated as irrecoverable low-temperature heat, by radiation from boiler, steam-pipes, engine, and as the heat which passes into the condenser water.

In the organism in general there is no distinction between the fixed parts of the mechanism and the working substance. The organism itself (its muscles, nerves, glands, etc.) *is* the working substance. Further, it is not quite certain that there is a necessary transformation of chemical energy into heat. The source of energy in the case of the warm-blooded animal is the chemical energy of the food substances and oxygen taken into its body. These chemical substances undergo transformations in the alimentary canal and in the metabolic tissues. The proteids of the food are broken down into animo-substances in the alimentary canal, and these animo-substances are synthesised into the specific proteids of the animal's body. Corresponding changes occur with the carbohydrates and fats ingested. These rearrangements of the molecular structure of the foodstuffs are the object of the processes of digestion and assimilation ; and when they are concluded, a certain proportion of the food taken into the body has become incorporated with, or has actually become a part of, the living tissues (muscles, nerves, etc.) of the animal body. This living substance, compounds of high chemical potential (proteids, carbohydrates, and fats) undergoes transformation into compounds of low chemical potential (water, carbon dioxide, and urea). There is a difference of energy, and this appears as mechanical energy, as the chemical energy required for glandular activity, and as heat.

We must not, however, conclude that this heat of the warm-blooded animal is comparable with the waste heat of the steam-engine. The homoiothermic animal maintains its body at a constant temperature, which is usually higher than that of the medium in which it lives, and this constancy of temperature obviously confers many advantages. Chemical reactions proceed with a velocity which varies with the temperature, so that in the warm-blooded animal the processes of life go on almost unaffected by changes in the medium. The animal exhibits complete activity throughout all the seasons of the year. It does not, or need not, hibernate, and it can live in climates which are widely different. We therefore find that the most widely-distributed groups of land-animals are the warm-blooded mammals and birds, while the largest and most cosmopolitan marine animals are the warm-blooded whales. Heat-production in the mammals and birds is therefore a direct object of the metabolism of the animal; it is a means whereby the latter acquires a more complete mastery over its environment. That it is not necessarily a step in the transformation of chemical into mechanical energy we see by considering the metabolism of the cold-blooded animals. In these poikilothermic organisms the body preserves the temperature of the medium. The temperature in such animals may be a degree, or a fraction of a degree, higher than that of the environment, but, in the absence of exact calorimetric experiments, we cannot say what proportion of the energy of the food of these animals passes into unavailable food energy. Probably it is a very small fraction of the whole, and we are thus justified in saying that in the cold-blooded animal chemical energy does not, to a significant extent, become transformed into heat. The result is, of course, that the vital processes in these organisms keep pace, so to speak, with the temperature of the environment, since the chemical reactions of their metabolism are affected by the external temperature. We find therefore that hibernation, the formation of resting stages, and a general slowing down of metabolic processes are more characteristic of the cold-blooded animal during the colder seasons than of the warm-blooded animal. The former has not that mastery over the environment attained by the mammal or bird.

The metabolism of the animal therefore resembles the energy process of the heat-engine only in the general way, that in both series of transformations chemical energy descends from a condition of high potential to a condition of low potential, transforming into mechanical energy in so doing, and thus performing work. In the heat-engine chemical energy transforms to heat, and then to mechanical energy, and of the total quantity transformed a certain large proportion suffers dissipation by conversion into low-temperature heat. In the animal organism chemical energy transforms directly to mechanical energy without passing through the phase of heat. If heat is produced it is because it is, in a way, available energy, inasmuch as it permits of the continuance of chemical reactions at a normal rate. The analogy of the animal with the heat-engine is therefore a false one. It suggests oxidation of the food-stuffs and heat production, whereas it is not at all certain that any significant proportion of the energy of the organism is the result of oxidation : many animal organisms indeed function in the entire absence of free oxygen. Further, the proportion of energy dissipated is always small compared with the heat-engine, and tends to vanish. The second law of thermodynamics does not, then, restrict the energy-transformations of the animal organism to the same extent that it restricts the energy-transformations of the physico-chemical mechanism.

The processes involved in the plant organism differ still more in their *direction* from those of a " purely physical " train. To see this clearly we must consider the imaginary mechanism known as a Carnot heat-engine.[1] This is a system in which we have (1) a heat-reservoir at a constant high temperature, (2) a refrigerator at a constant low temperature, and (3) a working substance which is a gas. Energy is drawn from the reservoir in the form of heat, and this heat expands the gas, doing work. The gas contracts, and its heat is then given up to the refrigerator. The work done is equal to the difference between the amount

[1] See appendix, p. 363.

of heat taken from the reservoir and the amount given to the refrigerator.

This series of operations is called a direct Carnot cycle. But the mechanism can be worked backwards. In this case heat passes from the refrigerator into the working substance, which was at a lower temperature. The working substance, or gas, is then compressed, as the result of which operation it is heated to just above the temperature of the reservoir. The heat it thus acquires is then given up to the reservoir.

In the direct Carnot cycle, therefore, energy passes from a state of high potential to a state of low potential and work is done *by* the mechanism. In the reversed Carnot cycle energy passes from a state of low potential to a state of high potential and work is done *on* the mechanism. The Carnot engine is thus perfectly reversible. No energy is dissipated in its working. It is, of course, a purely imaginary mechanism.

In the metabolism of the green plant carbon dioxide and water are taken into the tissues of the leaf and are transformed into starch. But the energy of the compounds, carbon dioxide and water, is much less than that of the same compounds when built up into starch. Energy must therefore be derived from some source, and this source is said to be the ether. Solar radiation is absorbed by the green leaf, and this energy is employed to produce the chemical transformation. Just how this is effected we do not positively know, in spite of much investigation. It is possible that formaldehyde is formed from carbon dioxide and water, polymerized, and then converted into starch. It is possible that the absorbed electromagnetic vibrations are converted into electricity in the chlorophyll bodies of the leaf, though when radiation is absorbed in physical experiments it is converted into heat. We do not know just what are the steps in the transformation, though it is clear that solar radiation is absorbed and that the chlorophyll of the leaf is instrumental in converting this energy of radiation into chemical potential energy. But the important thing to notice is, that we have here a process closely analogous to that of a reversed Carnot engine. Energy (that of the carbon dioxide and water) passes from a state of low potential to a state of high potential (that of the energy of starch), and work is done *on* the plant in producing this transformation.

Work is not done *by* the green plant. This statement is not, of course, quite rigidly true, for a certain amount of mechanical work is done by the plant. Flowers open and close ; tendrils may move and clasp other objects ; there is a circulation of protoplasm in the plant cells, and a circulation of sap in the vessels of stems, etc. Also work is done against gravity in raising the tissues of the plant above the soil, while work is also done by the roots in penetrating the soil. But when compared with the work done by radiation in producing the chemical transformations referred to above, these other expenditures of energy must be insignificant. Speaking generally, then, we may describe the green plant as a system in which available energy is accumulated in the form of chemical compounds of high potential. It is, further, a system in which energy becomes transformed without doing mechanical work, except to a trifling extent, and in which there is no formation of heat, or at least in which the quantity of heat dissipated is only perceptible during very restricted phases, is relatively small during the other phases, and tends to vanish.

Let us now combine the processes of plant and animal ; we start with the latter. In it we have a mechanism which does work. The source of its energy is the potential chemical energy of its food-stuffs, which latter reduce down to those substances known as proteids, fats, and carbohydrates. The energy-value of these compounds is considerable, that is to say, if they are burned in a stream of oxygen a large quantity of heat is obtained from their com-bustion. They are ingested by the animal, broken down chemically, and rearranged. The proteids eaten by the animal (say those of beef or mutton or wheat) are acted upon by the enzymes of the alimentary canal and are decomposed into their immediate constituents, animo-acids, and then other enzymes rearrange these animo-acids so as to form proteid again, but proteids of the same kinds as those characteristic of the tissues. This decomposition and re-synthesis is carried out also with respect to the fats and carbohydrates ingested. The result is that the food taken into the alimentary canal, or at least a part of it, is built up into the living substance of the animal's body. The energy expended upon these processes of digestion and assimilation is probably inconsiderable. During these processes the animal absorbs available chemical energy.

The energy thus taken into the animal is then transformed. The major part of it appears as mechani-cal energy—that of bodily movement, the movements of heart, lungs, blood, etc.—and heat. Some part of it becomes nervous energy, by which rather vague term we mean the energy involved in the propagation of nervous impulses. Some of it is used in glandular reactions, in the formation of the digestive juices, for instance. The most of it, however, transforms to mechanical energy and heat. Just how these energy transformations are effected we do not know. The heat is, of course, the result of chemical changes, oxidations, decompositions, or changes of the same kind as that of the dilution of sulphuric acid by water, but the mechanical energy appears to result directly from chemical change without the intermediation of heat. We shall return to this point in a later chapter, and content ourselves with saying here that the chemical compounds contained in the metabolic tissues of the animal body undergo transformation from a state of high to a state of low chemical potential, and that this difference of potential is represented by the work done and the heat generated. The proteid, fat, and carbohydrate of the tissues represent the condition of high potential; and the carbon dioxide, the water, and the urea, into which these substances are transformed, represent the condition of low potential.

Let us suppose a Carnot heat-engine in which the temperature of the reservoir of heat is (say) 120°C., and that of the refrigerator 50°C. The heat of the refrigerator can still be made a further source of energy by constituting it the heat reservoir of another Carnot engine which has a refrigerator at a temperature of 0°C. Our animal organism may be compared with a Carnot cycle; its energy reservoir is the proteid, fat, and carbohydrate ingested, and its refrigerator (or energy sink) is the carbon dioxide and urea excreted. Now the urea of the higher mammal becomes infected with certain bacteria, which convert it into ammonium carbonate. Another species of bacteria converts the ammonia into nitrite, and yet another turns the nitrite into nitrate. The main process of the animal is therefore combined with several subsidiary ones.

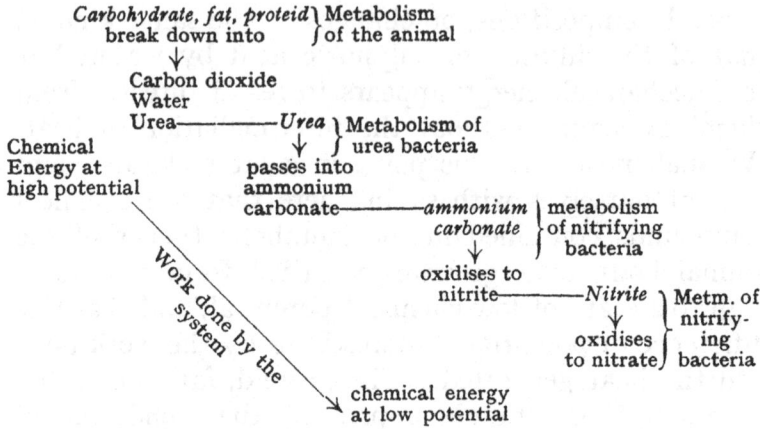

The arrows show that energy is descending the incline
indicated by a direct Carnot cycle. There is no more
work to be obtained from the carbon dioxide and water
excreted by the mammal, but more work can be obtained from the urea when it is used by bacteria, and
" ferments " to ammonia. Work can again be obtained
from the ammonia by bacteria, which convert it into
nitrite, and yet again from the nitrite by other bacteria,
which convert it into nitrate. The nitrate represents
the energy-zero so far as the organisms considered are
concerned,

Other nitrogenous residues are contained in the
urine of animals, and several other excretory products
may be formed. But in all these cases we can easily
find subsidiary energy-transformations effected by
bacteria, as in the above scheme. This, then, is the
positive, or direct half, of that reversible Carnot cycle
with which we are comparing life. In it energy falls
in potential (or intensity, or level), and in this fall of
potential transformations are produced—exhibit themselves, is perhaps a better way of putting it. We will
consider these transformations later ; in the meantime

it should be noted that in this fall of potential is a degradation of chemical energy. Compounds, carbon dioxide, water, and nitrate are produced which are chemically inert. It is no use to say that carbon dioxide may react with (say) glowing magnesium, water with metallic sodium, and nitrate with (say) glowing carbon. A condition of chemical equilibrium would result from purely inorganic becoming on our earth in which there was no metallic sodium or magnesium or incandescent carbon ; in which the metals would become inert oxides, and the carbon would become dioxide. The formation of these compounds represents a limit to energy-transformations. Note also that all these energy-transformations are conservative ; the total quantity remains unchanged throughout, and is the same at the end as at the beginning. But *entropy has been augmented : unavailable energy has increased at the expense of available energy.*

Consider now the indirect, or reversed, Carnot cycle. We begin with the inert matter, resulting from the metabolism of the animal, carbon dioxide, water, nitrate, and a few more mineral substances. We have the energy of solar radiation. By virtue of the living chlorophyll plastid in the cells of the green plant, this solar radiation uses the carbon dioxide and water as raw materials in the elaboration of starch. At the same time it absorbs nitrate, with some other inert mineral substances from the soil, and takes these into its tissues. The starch formed in the chlorophyll is converted into soluble sugar, which circulates through the vessels of the plant and is associated with the nitrogenous salt in the elaboration of proteid. Proteid, oils, fats and resins, and to a greater extent carbohydrates, are thus built up by the plant and *accumulate*, for mechanical work is not done by it, nor is heat dissipated—or at least these processes occur to an insignificant extent.

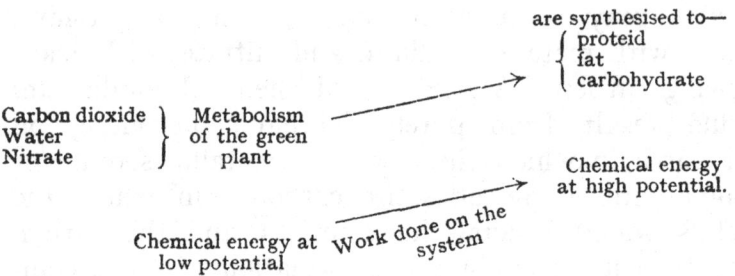

The "working substance" of our organic cycle has therefore returned to its original state.

We have considered the process of metabolism in two categories of organisms, the typical animal and the green plant, and we have combined these so as to obtain a picture of a reversible cycle of physico-chemical processes. When we speak of the "organism" in the most general sense, we mean that it exhibits these two modes of metabolism. This is, of course, not the case in any actual organism which we can investigate, or at least the typical modes of behaviour which characterise animal and plant life are not seen in any one individual. But we find that there is no absolute distinction between the two kingdoms. The plant may exhibit a mode of nutrition closely resembling that of the animal (as in the insectivorous plants), and it is possible that photo-synthetic process, in the general sense, may be present in the metabolism of some animals. Certain lower plants, the zoospores of algæ, exhibit movements identical in character with those of lower animals. At the base of both kingdoms are organisms, the

Peridinians, for instance, which have much of the structure of the animal (though cellulose is present in their skeleton), which possess motile organs, but which also possess a photo-synthetic apparatus, and exhibit the typical plant mode of nutrition. Further, there are symbiotic partnerships, that is, associations of plant and animal in one "individual" form (as, for instance, among the lower worms, Echinoderms, polyzoa, molluscs, and other groups of animals). In these cases green algal cells, capable of forming starch from carbon dioxide and water under the influence of light, become intercalated among the tissues of the animal. We find, also, that with regard to some fundamental characters, plant and animal display close similarities: the structure of the cell, for example, and the highly special mode of conjugation of the germ-nuclei in sexual reproduction. We must regard all the distinctive characters of the plant as represented in the animal and *vice versa*. Why they have become specialised in different directions is a question that we discuss later.

The organism, then, in so far as we regard it as a physico-chemical mechanism, as the theatre of energetic happenings, exhibits the following general characters :—

(1) It slowly accumulates available energy in the form of chemical compounds of high potential, work being done upon it.

(2) It liberates this energy in relatively rapid, controlled, "explosive reactions," transforming into movements carried out by a sensori-motor system of parts, work being done by it.

(3) In all these transformations the amount of energy which is dissipated is relatively small, and tends to vanish.

From the point of view, then, of energetic processes these are the characters of life, using the term in the general sense indicated above.[1]

Is there an absolute distinction between the organic mechanism and the inorganic one ?   Let us note, for the first time, that the actual physico-chemical transformations *themselves*, which we study in inorganic matter, are identical with those which we study in the organism.   Molecules of carbon dioxide, water, nitrate, sodium chloride, potassium chloride, phosphate, and so on, are just the same in inert matter as in the organism.   Chemical transformations, such as the hydrolysis of starch, the inversion of cane sugar, or the splitting of a neutral fat, are certainly just the same processes, whether we carry them out in the glass vessels of the laboratory, or observe them to proceed in the living tissues of the animal body.   The same molecular rearrangements, and the same transfers of energy, occur in both series of events.   This, however, is not the material of a distinction : what we have to find is, whether the *direction* of a group of physico-chemical reactions is the same in the organism and in a series of inorganic processes.

Let us return to the Carnot cycle.   This is a series of operations which occur in an imaginary mechanism in such a manner that the whole series can be easily reversed.   Heat is supplied to the imaginary engine, which then performs work and yields up its heat to a refrigerator.   Work is then performed on the engine, which thereupon takes heat from the refrigerator and returns it to the source.   The work done *by* the engine in the direct cycle is equal to the work done *on* it in

[1] This is, of course, the argument of part of Chapter II. of Bergson's *Creative Evolution*.   The reader will not find the essential differences between plants and animals stated so clearly anywhere else in biological literature.

the indirect cycle. The heat taken from the source and given to the refrigerator in the direct cycle is equal to the heat taken from the refrigerator and given to the source in the indirect cycle. But it is a purely imaginary mechanism, and all experience shows not only that it has not been realised in practice, but that it cannot so be realised. If it could be realised, we should show that the second law of thermo-dynamics is not physically true.

Do the energy processes of life realise such a perfectly reversible cycle of operations ? In order to answer this question we must consider the fate of the energy which is absorbed in the plant metabolic cycle, and that which is given out in the animal one. Does *all* the energy of solar radiation which is absorbed by the plant pass into the form of the potential chemical energy of the carbohydrates and other substances manufactured ? Does any of the energy of the animal which results from the metabolism of its body pass into the unavailable form—that is, into a form in which it cannot be utilised by other organisms ? That is to say, is energy dissipated by the organism ?

Undoubtedly it is to some extent, but to a far less extent than in the inorganic train of processes. Some of the energy of solar radiation absorbed by the plant must become transformed, by the friction of whatever movements occur, into low-temperature heat, and some quantity of heat, however small, is generated by the metabolism of the plant. Again, some of the heat of the warm-blooded animal must be radiated into space, or conducted away from its body ; and this energy becomes dissipated—let us assume, at least, that it is so dissipated in the physical sense. Probably also some quantity of heat is generated by the metabolism of the cold-blooded animal, though this must be a very small proportion of the total energy transformed. We see, then, that the distinction is one of degree, though the difference between inorganic and organic energetic processes is very great in this respect ; so great that we must regard it as constituting a fundamental difference, and as indicative of the limitation of the second law when extended to the functioning of the organism.

But we have also to consider the effect of the work done by the organism. We consider the nature and meaning of the evolutionary process in a later chapter, but in the meantime we may state this thesis : that the process of evolution leads up to man and his activity. It *leads*, if we regard the process as a *directed* one ; but even if we regard it as a fortuitous process we still find that man, far more than any other organism, is the result of it. All the facts of biology and history show that man dominates the organic world, plant or animal ; that the whole trend of his activity is to eliminate whatever organisms are inimical, and to foster those that are useful. Already, during the brief period of his rational activity, the wolf has disappeared from civilised lands while the dog has been produced. Species after species of hostile or harmful organisms have been, or are being, destroyed or changed, while numerous other species have been preserved and altered for his benefit. In the future we see an organic world subservient to him either entirely or to an enormous extent.

So also in the inorganic world. Rivers which formerly rushed down through rapids, dissipating their energy of movement in waste irrecoverable heat, now pour through turbines and water wheels, generating electricity and accumulating available energy. Winds which " naturally " dissipated their mechanical energy in waste heat now propel ships and windmills. Tides, with their incredibly great mechanical energy, now simply warm up the crust of the earth by an infinitesimal fraction of a degree daily, and produce heat which at once radiates into space. Who doubts that by and by this energy too will become accumulated for human use? Multitudes of chemical reactions were potential, so to speak, in the molecules of petroleum, while the energy which might have produced them ran to waste. But under human activity this energy became directed and made to produce chemical reactions formerly existing only in their possibility, and all the substances of modern organic chemistry came into existence.

The energy, then, of human activity has been directed towards averting or retarding the progress towards dissipation, or irrecoverable waste, of cosmic energy— that of the sun's radiation, and of the motions of earth and moon. Human activity has accumulated available energy. The difference of water-level between Niagara and the rapids below represents available mechanical energy. A few years ago an enormous quantity of this energy became irredeemably lost in waste heat every twenty-four hours: now it remains available for work; and this quantity of work retained is enormously greater than is the human energy which was expended on erecting the water-power installation there.

The processes studied by physics and chemistry are therefore irreversible ones. We can conceive a perfectly reversible process, as in the Carnot heat-engine, but this is a purely intellectual conception, formed as the limit to a series of operations which approximate closer and closer to an ideal reversibility. It is a conception that has no physical reality—a

F

guide to reasoning only. On the other hand we see that all naturally occurring physical processes are irreversible and in their sum tend to complete degradation of energy. Mechanistic biology isolates physico-chemical processes in the functioning of the organism, and sees that they conform to the law of dissipation, as well as to that of the conservation of energy.

Yet the organism as a whole, that is, life as a whole, on the earth, does not conform to the law of dissipation. That which is true of the isolated processes into which physiology decomposes life is not true of life. In all inorganic happenings energy becomes unavailable for the performance of work. Solar radiation falling on sea and land fritters itself away in waste irrecoverable heat, but falling on the green plant accumulates in the form of available chemical energy. The total result of life on the earth in the past has been the accumulation of enormous stores of energy in the shape of coal and other substances. By its agency degradation has been retarded. Whenever, says Bergson, energy descends the incline indicated by Carnot's law, and where a cause of inverse direction can retard the descent, there we have life.

CHAPTER III

THE ACTIVITIES OF THE ORGANISM

THE rather lengthy discussion of the last chapter was necessary in order to show just how far the principles of energetics established by the physicists applied to the organism. We have seen that the first law of thermodynamics does so apply with all its exclusiveness. The more carefully a physiological experiment is made; the more closely do its results correspond with those which theory demands. It is true that relatively few experimental investigations can be controlled in this way, but in those that can be checked by calculation (as, for instance, in the well-known calorimetric experiments) everything tends to show that precisely the same quantities of matter and energy enter the body of an organism in the form of food-stuff, that leave it as radiated and conducted heat, as work done, and as the potential chemical energy of the excretions. Even when we are unable (as in most investigations) to apply the test of correspondence with theory, we have the conviction that the law of conservation holds with all its strictness.

Then, whenever it was possible to apply the methods of chemistry and physics to the study of the organism, it was seen that the processes at work were chemical and physical. The substance of the living body was seen to consist of a large (though limited) number of chemical compounds, differing mainly from those which exist in inorganic nature in their greater complexity It was also seen that physico-chemical reactions occurred in living substance analogous with, or quite similar to, those which could be studied in non-living substance.   The conclusion, then, was irresistible that the life of the organism was merely a phase in the evolution of matter and energy, and differed in no essential respect from the physico-chemical activities that could be observed in the non-living world.

These conclusions were stated so well by Huxley in his famous lecture on " The physical basis of life," over forty years ago, that all subsequent utterances have been merely reiterations of this thesis in a less perfect form.   The existence of the matter of life, Huxley said, depended on the pre-existence of certain chemical compounds—carbonic acid, water, and ammonia.   Withdraw any one of them from the world and vital phenomena come to an end.   They are the antecedents of vegetable protoplasm, just as the latter is the antecedent of animal protoplasm.   They are all lifeless substances, but when brought together under certain conditions they give rise to the complex body called protoplasm ; and this protoplasm exhibits the phenomena of life.   There is no apparent break in the series of increasingly complex compounds between water, carbon dioxide, and ammonia, on the one hand, and protoplasm on the other.   We decide to call differen tkinds of matter carbon, oxygen, hydrogen, and nitrogen and to speak of their activities as their physico-chemical properties.   Why, then, should we speak otherwise of the activities of the substance protoplasm ?

" When hydrogen and oxygen are mixed in certain proportions and an electric spark is passed through them they disappear, and a quantity of water, equal in weight to the sum of their weights, appears in their place. There is not the slightest parity between the passive and active powers of the water and those of the oxygen and hydrogen that have given rise to it. . . . We call these and many other phenomena, the properties of water, and we do not hesitate to believe that in some way they result from the properties of the component elements of the water. We do not assume that a something called " aquosity " entered into and took possession of the oxide of hydrogen as soon as it was formed and guided the aqueous particles to their places in the facets of the crystal, or among the leaflets of the hoar frost."

" Is the case in any way changed when carbonic acid, water, and ammonia disappear, and in their place, under the influence of pre-existing protoplasm, an equivalent weight of the matter of life makes its appearance ? "

" It is true that there is no sort of parity between the properties of the components and the properties of the resultant. But neither was there in the case of water. It is also true that the influence of pre-existing protoplasm is something quite unintelligible. But does anyone quite understand the *modus operandi* of an electric spark which traverses a mixture of oxygen and hydrogen ? What justification is there, then, for the assumption of the existence in the living matter of a something which has no representative or correlative in the non-living matter which gave rise to it ? "

All the investigations of over forty years leave nothing to be added to this statement of what, in Huxley's days, was called materialistic biology. It was a very unpopular statement to make then, but it has become rather fashionable now. Let the reader compare it with all that has been spoken and written since 1869, even with the utterances of the British Association of the year 1912, and he will find that it expresses the point of view of mechanistic biology far better than all the subsequent restatements. The only difference he will find is that the latter have become (as William James has said about academic philosophies), rather shop-soiled. They have been reached down and shown so often to the enquiring public, that each display has taken away something of their freshness.

Now Huxley's example leads up so well to the consideration of the differences between the chemical activities of the organism and those of inorganic matter that we may consider it in some detail. What, then, *is* the difference between the explosion of a mixture of oxygen and hydrogen, and the photo-synthesis of starch by the green plant ?

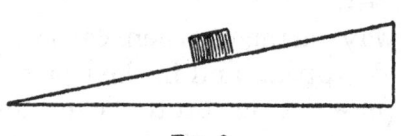

FIG. 8.

In the case of the synthesis of water we have an example of an exothermic chemical reaction. We are to think of the mixture of oxygen and hydrogen as existing in a condition of " false equilibrium." It may be compared with a weight resting on an inclined plane.

Suppose that the plane is a sheet of smoothly polished glass, and that the weight is a smooth block of glass. By canting the plane more and more an angle will be found at which the slightest push starts the weight sliding down. Now in the case of the explosive mixture of oxygen and hydrogen we have a chemical analogue. Either the gases do not combine at all at the ordinary temperature or they combine " infinitely slowly."

But the slightest impulse, an electric spark requiring an almost infinitesimally small quantity of energy, starts the combination of the gases, and this continues until all is changed into water vapour. In this reaction a large quantity of energy is liberated in the form of heat. This heat becomes transformed into the kinetic energy of the water particles which condense from the steam formed in the explosion, and these particles assume the temperature of their surroundings. The energy which was potential in the explosive mixture, and which was capable of doing work, still exists as the kinetic energy of the water formed, but it has become unavailable for any natural process of work.

We have seen what is the general character of the reaction series in the course of which carbon dioxide and water become starch ; and then this, becoming first soluble, and becoming associated with the ammonia or nitrate taken into the plant, becomes protoplasm. It is a reaction which differs from that just described, in that available energy becomes absorbed and accumulated, and retains the power of doing work. It is not a reaction which can be initiated by an infinitesimal stimulus, but one in which just as much energy is required in order that it may happen as is represented in the energy which becomes potential in the living substance generated. The first reaction is one which may take place *by itself* ; [1] the other is one which requires a compensatory energy-transformation in order that it may happen. In the first reaction energy is dissipated ; in the second one it is accumulated.

---

[1] It is no use saying that apart from the electric spark the combination would not take place, for we do not know that the O and H of the mixture do not combine *very* slowly, molecule by molecule, so to speak. At all events there is no functionality between the infinitesimal quantity of energy supplied by the spark, and the energy which becomes kinetic in the explosion.

We are thus led to the consideration of the second principle of energetics and its limitations, but before entering upon this discussion we must consider the nature of the activities of the organism.

By the term " metabolism " we understand the totality of the physico-chemical changes which occur in the living substance of the organism. In physiological writings we usually find that two categories of metabolic changes are described : (1) anabolic processes, in the course of which simple chemical compounds possessing relatively little energy are built up into much more complex substances, containing a relatively large quantity of available energy, and therefore capable of doing work. The transformations constituting an anabolic change must be accompanied by corresponding compensatory energy-transformations, to account for the energy which becomes potential in the substances formed. The formation of starch from carbon dioxide and water, by the green plant, is such an anabolic change, and the compensatory energy-transformation is the absorption of radiation from the ether by the cells of the plant. A further anabolic change in the plant organism is the formation of amido-substances from the ammonia or nitrate absorbed from the soil, and from the soluble carbohydrates formed from the starch manufactured in the green cells.

The typical activities of the chlorophyll-containing organism are of this nature ; they are anabolic. The organism may be a green land-plant ; a marine green, red, or brown alga ; a yellow-green diatom, a yellow, green, red, or brown peridinian or other holophytic protozoan ; an ascidian, mollusc, echinoderm, polyzoan, worm, or coral containing " symbiotic algae " (that is the chlorophyll - containing cells of some plant organism which have become associated with the animal and incorporated in its tissues). In all these cases the presence of this chlorophyllian substance confers on the organism the power of effecting the compensatory energy-transformation, by the aid of which carbon dioxide and water are built up into starch. What this transformation is, and what are the steps by which the carbon dioxide and water become carbohydrate we do not exactly know. Solar radiation impinging upon an inorganic substance is partly reflected and partly absorbed. The absorbed fraction may become transformed in such a way as to render the substance phosphorescent, or it may transform into chemical energy, as when light impinges on a photographic plate, but as a general rule it is transformed into heat. In the green plant, however, the transformation of radiation into heat does not occur—at least the heating is very small—and it passes directly or indirectly into the potential chemical energy of the starch which is synthesised. We must regard this power of absorbing radiation and utilising it in compensatory transformations as a general character of protoplasm. It is true that it is now specialised in the cells containing the chlorophyll bodies, but there are indications that it may be present in the tissues of the animal devoid of chlorophyll.

Other anabolic transformations occur in the animal. The food-stuffs which are absorbed from the intestine are substances which have undergone dissociations, the nature of which is such as to render them capable of absorption and of reconstruction. These anabolic changes in the higher animal are exceptional, and their usefulness lies in the fact that by their means substances become capable of being transported by the tissue fluids of the body.

(2) Katabolic changes in the animal body correspond in their frequency of occurrence to the anabolic changes of the plant organism. In them complex chemical substances undergo transformation into relatively simple substances, and the contained energy at the same time undergoes a parallel transformation, passing into the form of heat and mechanical energy, while a fraction becomes dissipated. Food-stuffs taken into the alimentary canal break down in this way, but to a very limited extent. Proteids undergo dissociation or decomposition into amido-substances, while fats are dissociated into fatty acids and glycerine. Doubtless energy is dissipated in these processes, serving no other purpose but to heat the contents of the alimentary canal, but this energy-transformation has not been worked out very completely and it is a question whether, given a healthy animal and perfect food-stuffs, any energy would necessarily be lost during the digestive processes. The reactions involved in the latter do not belong to the category of chemical changes proceeding from the complex to the simple, with a liberation of energy ; but appear to involve rather a rearrangement of the constituents of a complex molecule, a process in which the contained energy need not undergo change in quantity. These processes involve the action of *enzymes*.

Enzymes play a great part in modern physiological theory and we must consider them in detail. Let us attach a concrete meaning to the general notion of enzyme-activity by considering the phenomena known as *catalysis*. The metal platinum can be brought into a very fine stage of division when it is known as platinum black. In this condition it brings about reactions in chemical mixtures or substances which would not otherwise occur : a mixture of oxygen and hydrogen explodes when brought in contact with platinum black, and a mixture of coal gas and air inflames, a reaction which is made use of in the little gas-lighting apparatus which most people have seen. If, again, a powerful electric current be passed between platinum wires which are a little distance apart, and are immersed in water, the metal becomes torn away from the points of the wire in the form of an impalpable powder, colloidal platinum. The liquid containing this colloid then has the power of setting up chemical changes in other substances, changes which would not otherwise occur, or, at least, would occur very slowly.

In general such catalysts, platinum black or colloidal platinum for instance, have the following characters : (1) a small quantity is sufficient to cause change in a large (theoretically an infinite) quantity of the substance acted upon ; (2) the nature and quantity of the catalyst remain at the end the same, as at the beginning of the reaction ; (3) a catalyst does not start a reaction in any other substance or sub-stances, it can only influence the rate at which this reaction may occur : apparently it does, in some cases, start a reaction, but in such cases we suppose that the latter proceeds so slowly as to be imper-ceptible ; (4) the final state of the reaction is not affected by the catalyst ; it depends only on the nature of the interacting substance or substances ; (5) the final state is not affected either by the nature or quantity of the catalyst : it is the same if we employ different catalysts, or a large or small quantity of the same catalyst. Finally, it appears that the phenomena of catalysis are universal : " There is probably no kind of chemical reaction," says Ostwald, " which cannot be influenced catalytically, and there is no substance, element, or compound which cannot act as a catalyser."[1]

Enzymes, then, are agents which are produced by the organism, and which act by influencing (accelerating or retarding) chemical reactions. An enzyme, as such, need not exist in a tissue; it is there as a *zymogen*, a substance which may become an enzyme when required. An enzyme need not be active : it may be necessary that it should be " activated " by a *kinase*, another substance produced at the same time. Associated with many enzymes are *anti-enzymes*, substances which undo what their corresponding enzymes have done. Finally some, perhaps most, enzymes are reversible, that is, if they produce a change in a certain substance they can also produce the opposite kind of change : the meaning of this will become clearer a little later on. We have spoken of enzymes as " agents " or " substances," but it is not at all certain that they are definite chemical compounds. In the preparation of an enzyme what the bio-chemist obtains is a liquid, a glycerine or other extract *which possesses catalytic properties*. An actual catalytic substance, like platinum black, cannot be obtained from this liquid. A white powder may be obtained, but this usually proves to be proteid in composition ; it is not the actual enzyme itself but is the impurity associated with the latter. Now the very great number of enzymes " isolated " by the physiologists has rather destroyed the original simplicity of the idea of enzyme activity and suggests a parallel statement to that made by Ostwald about catalysts : any tissue substance may influence the reactions that may possibly occur in

[1] A statement of interest in view of the enormous number of " ferments " or enzymes discovered by physiologists. It would appear that any tissue in any organism is capable of yielding an enzyme to modern investigation.

other tissue substances. But while pure chemistry has to deal with definitely known chemical compounds in the phenomena of catalysis, this cannot be said to be the case with physiology in dealing with enzymes. Reasoning by analogy, we may say that it is probable that enzymes are definite proteids, or chemical substances allied to these, but this has not been clearly demonstrated, and it is possible that the phenomena of enzyme activity may belong to some other category of energy-transformations.

However this may be, the conception is a useful one in describing the reactions of the organism, and it may be illustrated by considering the digestion and absorption of fat in the mammalian intestine, a process which appears to be better known than that of proteid digestion. A neutral fat consists of an acid radicle, oleic, palmitic or stearic acids, for instance, united with glycerine. The action of the pancreatic or intestinal enzymes is to dissociate this fatty salt. Let us write the formula of the latter as $G F$, G being the glycerine base, and F the fatty acid ; then

$$G F \rightleftarrows G + F$$

which means that the enzyme can cause the neutral fat to dissociate into glycerine and fatty acid. This action will go on until a state of equilibrium is attained, in which there is a certain quantity of each of the radicles, and a certain quantity of unchanged neutral fat, the ratio of all these to each other depending on various things. When this state of equilibrium is attained the enzyme does indeed go on splitting up more neutral fat, but it is a *reversible* enzyme, and it also causes the glycerine and fatty acid already split up to recombine, forming neutral fat. A condition is, therefore, reached in which the composition of the mixture remains constant.

Now there is dissociated fat in the intestine after a meal, but there is only neutral fat in the wall of the intestine. The fat itself cannot pass through the cells forming the intestinal wall, but the glycerine and fatty acid into which it is dissociated can so pass, since they are soluble in the liquids of the intestine. We suppose that the cells of the wall of the intestine also contain the fat-splitting ferment; this ferment in the cells acts on the glycerine and fatty acid immediately they enter and recombines these radicles again into neutral fat, the above equation now reading from right to left. But after a time this reaction in the cells will also begin to reverse, for the enzyme will begin to split up the synthesised neutral fat when the state of chemical equilibrium in the new conditions is attained. Fatty acid and glycerine will then diffuse out from the cells into the adjacent lymph stream or blood stream— perhaps neutral fat will also pass from the cells into these liquids, we are not sure. At all events the lymph and blood after a meal containing much fat are crowded with minute fat globules. But why are there no fatty acids or glycerine in the blood, for the latter also contains lipase (the fat-splitting enzyme)? The explanation is, apparently, that either an anti-enzyme is produced, or that the enzyme passes into a *zymoid* condition. Why also does fat accumulate in the tissues? Here, again, the activity of the enzyme, which from other considerations we may regard as being universally present almost everywhere in the body, must be supposed to be arrested by some means.

The conception of a catalytic agent, such as we can study in pure chemistry, thus carries us a long way in our description of the processes of digestion, absorption, and assimilation. We have applied it to the case of fat-digestion, but very much the same general scheme might also apply to many other processes in the body. Obviously it enables us to describe these processes in terms of physico-chemical reactions, but we cannot fail to see that ultimately we are compelled to assume the existence of reactions which were not included in the original conception—the activation of the enzyme at the proper moment by the kinase, the operation of the anti-enzyme, and the passage of the enzyme into the zymoid. Just why these things happen as they do we do not know, yet the whole problem becomes shifted on to these reactions.

In the same way we apply the purely physical processes of the osmosis and diffusion of liquids to the circulation of substances in the animal body. The nature of these processes will probably be familiar to the reader, nevertheless it may be useful to remind him that by diffusion we understand the passage of a liquid, containing some substance in solution, through a membrane ; and by osmosis the passage of a solvent (but not of the substance dissolved in it) through a " semi-permeable membrane." The molecules of the solvent (water, for instance) pass through the membrane (the wall of a capillary, or lymphatic vessel), but the molecules of the substance (salt, for instance) dissolved in the solvent do not pass. Let us suppose that a strong solution of common salt in water is injected into the blood stream : what happens is that osmosis takes place, the water in the surrounding lymph spaces passing into the blood stream because the concentration of salt there is greater than it is in the lymph. While this is happening, the capillary walls are acting as semi-permeable membranes, allowing the molecules of water to pass through but not the molecules of salt.

Very soon, however, the process of osmosis becomes succeeded by one of diffusion, and the salt molecules pass through the capillary wall into the lymph and are excreted.

Undoubtedly the purely physical processes of diffusion and osmosis occur all over the animal body and are the means whereby food-materials, secretory, and excretory substances are transported from blood to lymph, or *vice versa*, from lymph to cell substance or to glandular cavities, and so on. But it is also the case that in very many processes the activity of the cells

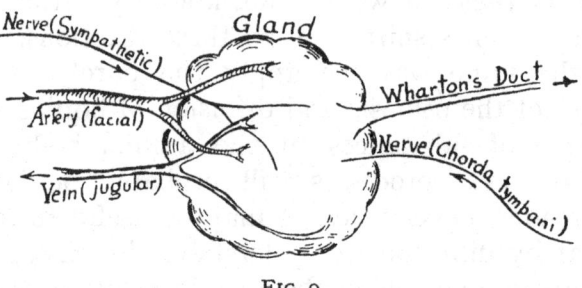

Fig 9.

themselves plays an important part. It may even be the case that a particular process, after all physical agencies are taken into account, reduces down to this action of the cells. To understand this we must consider the mode of working of some well-known organ, and the best possible example of such an organ, considered as a mechanism, is that of the sub-maxillary salivary gland of the mammal.

What, then, is this mechanism and how does it act ? The gland is a compound tubular one, its internal cavity being prolonged into the duct which opens into the mouth. The saliva prepared in the gland issues from this duct. Blood is carried to the gland by twigs of the facial artery, and, after circulating through it,

is carried away by factors of the jugular vein. Two
nerves supply the gland : one is the chorda tympani,
a branch of a cranial nerve, and the other is a sympathetic nerve. Lymph also leaves the gland by a
little vessel.

Now suppose we have laid bare all this mechanism
in a living animal and make experiments upon it.
If we stimulate the chorda tympani there is a copious
flow of thin watery saliva, but if we stimulate the
sympathetic there is a less copious flow of thick viscid
saliva. Why is this? We find on closer analysis
that the chorda contains fibres which dilate the small
arteries so that there is an increased flow of blood
through the gland; but that, on the other hand, the
sympathetic contains fibres which constrict the
arteries, thus leading to a reduced flow of blood. This
accounts for the fact that " chorda-saliva " is abundant
and thin, while " sympathetic-saliva " is scarce and
thick. It was thought at one time that the chorda
contained fibres which stimulated the gland to produce
watery saliva, while the sympathetic contained fibres
which stimulated it to produce mucid saliva. This,
however, is not the case. Both nerves contain the
same kind of secretory fibres : their other fibres differ
mainly in that they act differently on the arteries.

It might be the case—indeed it was at one time
thought that it was the case—that secretion of saliva
was simply a matter of blood-flow : an abundant
arterial circulation gave rise to abundant saliva, a
sparse flow to a sparse saliva. Undoubtedly the
secretion depends on blood supply, but not *solely*.
If it did, then the whole process might be conceived to
be a very simple mechanical one—filtration or diffusion
of the saliva from the blood stream through the thin
walls of the blood vessels, and the walls of the tubules

G

into the cavity of the gland. If this were the case, then the liquid in the gland would be the same in composition and concentration as the liquid part of the blood—the plasma. But it is really different in composition and it is not so concentrated. Now osmotic pressure—on the action of which so much is based—cannot help us, for the liquid in the gland is *less* concentrated than that in the blood vessels, so that water ought to pass from gland to blood instead of from blood into gland. Again, if we tie the duct, so that the saliva cannot escape, secretion still goes on, though the hydrostatic pressure of saliva in the cavity of the gland may be considerably greater than that of the liquid in the blood vessels. Yet again, if we stop the blood flow by tying the artery, secretion of saliva may still go on for a time.

Therefore the only physical agencies we can think of do not explain the secretion. The latter is actually the work of the individual cells, stimulated by the nerves. If the volume of the gland be measured just while it is being stimulated to secrete, it will be found that the organ becomes *smaller*, yet while it is being stimulated the blood-vessels are being dilated so that the volume of the whole structure ought to become greater. Obviously part of the substance of the gland is being emptied out through its duct as the secretion.

If we examine the cells of the gland in various states we see clearly that granules of some material, different in nature from the substance of the protoplasm itself, are being formed within them. Evidently these granules swell up during secretion and discharge their contents into the ducts. Further changes in the characters of the cell-substance, and in the nucleus, can be observed, and all these indicate that the protoplasm of the cells, as the result of stimulation, elaborates certain substances ; that these substances are then washed out, so to speak, into the duct by the withdrawal of water from the cell ; and that thereafter the cell absorbs fresh nutritive material from the lymph which exudes from the blood vessels, along with water. The distinctive part of the whole train of processes is, then, this elaboration of material by the cells themselves ; while the concomitant changes in the calibre of the blood vessels and in the flow of blood and lymph are subsidiary ones. In the process of secretion of saliva energy is absorbed from the chemical substances of the blood to bring about the passage of water from a region of high to a region of low osmotic pressure ; oxygen and nitrogen, with other elements of course, are withdrawn from the arterial blood stream for the purpose of the secretion, and carbon dioxide and other substances are given off to the venous blood and lymph.

The problem thus is pushed back from the mechanical events occurring in the nervous and circulatory processes, to the physico-chemical ones occurring in the cells of the gland tubules ; and it thus becomes much more obscure. It is true that we can formulate a hypothesis which describes, in a kind of way, these intra-cellular metabolic changes, in terms of physico-chemical reactions, and, without doubt, reactions of this kind must occur within the cell. But if we could test any such hypothesis as easily as the mechanical ones suggested, should we find it any more self-sufficient ? [1]

---

[1] We have not referred to " psychical secretion." If we smell some very savoury substance our " mouth waters," that is, secretion of saliva occurs. If we even see some such substance the same secretion occurs. All this is clear and can be " explained " mechanistically : the stimulation of the olfactory or visual organs begins a kind of reflex process. But if we even *think* about some very savoury morsel saliva may be secreted. We must

Irritability and contractility are general properties of the organism. These properties are illustrated by the irritability of an *Amœba* or *Paramœcium* to stimuli of many kinds ; by the movements of the pseudopodia of the former animal, or of the cilia of the latter ; by the nervous irritability of the higher animal, and the contraction of its muscles when they are stimulated. They are among the fundamental properties or functions of living protoplasm, and their study is of paramount interest, and carries us to the very centre of the problem of the activities of the organism. Naturally physiologists have never ceased to attempt to describe irritability and contractility in terms of physics, but though we may be quite certain that the things that do occur in these phenomena are controlled physico-chemical reactions, it must be remembered that what we positively know about their precise nature is exceedingly little.

What is the nature of a nervous impulse ? When a receptor organ is stimulated, as, for instance, when light impinges on the cone cells of the retina, or when the nerve-endings in a " heat-spot " in the skin are warmed, or when the wires conveying an electric current are laid on a naked nerve, an impulse is set up in the nerve proceeding from the place stimulated, and we must suppose that approximately the same amount of energy moves along the nerve as was communicated to the receptor or the nerve itself by a stimulus of minimal strength. How does it so move ?

suppose now that our consciousness, something which has nothing to do, it must be noted, with energy—changes in the body, can react on the body. If we show a dog an attractive bone it will secrete saliva; if we show it again and again, the same thing occurs. But after certain such trials the dog will realise that he is being played with, and the exhibition of the bone no longer evokes a flow of secretion. Why is this ? The whole process has now become more mysterious than ever.

Several facts of capital importance result from the experimental work. (1) The impulse travels with a velocity variable within certain limits, say from 8 to 30 metres per second ; (2) it travels faster if the temperature is raised (up to a certain limit) ; (3) it is difficult to demonstrate that the passage of this impulse is accompanied by definite chemical changes in the nerve substance : it is stated that carbon dioxide is produced, but this is not certainly proved ; (4) an electric current is produced in the nerve as the result of stimulation ; (5) no heat is produced, or at least the rise of temperature, if it occurs, is less than $0.0002°$ C.

Thus it is quite certain that physical changes accompany the propagation of the nerve-impulse, for the latter has a certain velocity, which depends on the temperature, and an electric change also occurs in the substance of the nerve. Is this electric change the actual nerve impulse ? It is hardly likely, since the velocity of the impulse is very much less than that of the propagation of an electric change through a conductor ; besides, the passage of the impulse is not accompanied by a measurable heat evolution, although the flow of electricity along a poor conductor must generate heat and dissipate energy. Is it a chemical change ? Then we should be able to observe metabolism in the nerve substance—that is if the energy-change is a thermodynamic one—while it is not at all certain that metabolic changes do occur. Nevertheless it seems probable that a physico-chemical change is actually propagated when we consider the chemical specialisation of the substance of the axis-cylinder of the nerve. Now the velocity of propagation of the nervous impulse is of the same order of magnitude as that of an explosive change in chemical substances (using the term " explosion " to connote chemical disintegrations rather than combustions). If we imagine a long rod of dynamite, or picric acid, or a long strand of loosely-packed gun-cotton to be exploded by percussion at one end, then a transmission of the chemical disintegration of any of these substances will pass along the rod, etc., with a velocity which will certainly vary with the physical condition of the material. It would be a high velocity in a rod of dynamite, or fused picric acid, but a lower velocity in a loosely aggregated strand of gun-cotton, or a trail of picric acid powder. Is this what happens in the nerve when an impulse travels along it? Obviously not, since the substance of the nerve is not altered appreciably, while that of the explosive substance passes into other chemical phases. We might imagine, then, such a change in the nerve fibrils as that of a reversible transformation of some chemical constituent :—

$$
\begin{array}{ccccc}
 & & (2) & (1) & \\
\hline
: a+b : & a+b : & a+b : & a+b : & a+b : \\
: \downarrow\uparrow : & \uparrow\downarrow : & \uparrow\downarrow : & \uparrow\downarrow : & \uparrow\downarrow : \\
: c+d : & c+d : & c+d : & c+d : & c+d : \\
\hline
\end{array}
$$

Let us imagine the substance of the fibril to be composed of, or at least to contain, the substances $a+b$ which dissociate reversibly into the substances $c+d$. At any moment, and in any particular physical state, as much of a and b pass into c and d as c and d pass into a and b. There will be equilibrium. But now let a stimulus alter the physical conditions : prior to the stimulus the phase was $a_m+b_n = c_p+d_r$,—the suffixes m, n, p, r, denoting the concentrations of a, b, c, and d—but after the stimulus the phase may be $a_{m1}+b_{n1}=c_{p1}+d_{r1}$. Now the element of the nerve substance (1) forms a system with the element (2).   The condition in (2) is $a_m + b_n = c_p + d_r$, and that of (1) $a_{m1} + b_{n1} = c_{p1} + d_{r1}$, but these two together now fall into a new state of equilibrium and this is transmitted along the whole nerve-fibril with a velocity which belongs to the order of magnitude of that of chemical changes.   If the stimulus remains constant (a constant electric current for instance), the new condition of equilibrium will be established throughout the whole length of the fibril and the nervous impulse will be a momentary one (as it is in this case).   But if the stimulus is an intermittent one (an interrupted electric current, light-vibration, sound-vibrations), then in the intervals the former condition of equilibrium will become re-established and the nervous impulse will be intermittent (as it is).   There would be no work done on the whole in the changes, except that done by the transmission of the changed state of equilibrium to the substance of the effector organ in which the nerve-fibril terminates —the substance of a muscle fibre, or the cell of a secretory gland, for instances.   There would, probably, be a certain dissipation of energy as in the case of the propagation of an electric impulse through a poor conductor, but all our knowledge of the chemistry of the nerve fibre points to this amount of dissipation as tending to vanish.

Something analogous to this may be expected to take place in a muscle fibre when it contracts ; except that, of course, energy is transformed in this case. What precisely does happen we do not know and at the present time no physico-chemical hypothesis of the nature of muscular contraction exactly describes all that can be observed to take place.   Certain positive results have, of course, been obtained by chemical and physical investigation of the contracting muscle : carbon dioxide is given off to the lymph and blood stream, and the amount of this is increased when an increased amount of work is done by the muscle ; heat is produced and this too increases with the work performed ; glycogen is used up, and lactic acid is produced ; finally oxygen is required, and more oxygen is required by an actively contracting muscle than by a quiescent one. Now the obvious hypothesis correlating all these facts is that the muscle substance is oxidised, and that the heat so produced is transformed into mechanical energy. " We must assume," says a recent book on physiology, " that there is some mechanism in the muscle by means of which the energy liberated during the mechanical change is utilised in causing movement, somewhat in the same way as the heat energy developed in a gas-engine is converted by a mechanism into mechanical movement."

Now, must we assume anything of the kind ? To begin with, life goes on, and mechanical energy is produced in many organisms living in a medium which contains no oxygen. Anaerobic organisms are fairly well known, and we cannot suppose that in them energy is generated by the combustion of tissue substance in the inspired oxygen. A muscle removed from a cold-blooded animal will continue to contract in an atmosphere containing no oxygen, and it will continue to produce carbon dioxide. It is true that the contractions soon cease, even after continued stimulation under conditions excluding the fatigue of the muscle, but do the contractions cease *because* the oxygen supply is cut off, or because the muscle dies in these conditions ? We know that some complex chemical substance is disintegrated during contraction and that mechanical energy and heat are produced and that carbon dioxide is also produced. We know that the carbon contained in the latter gas corresponds roughly with the carbon contained in the muscle substance which undergoes disintegration, but does all this justify us in saying that this substance is oxidised in order that its potential chemical energy may be transformed into mechanical energy? Obviously not, since we might equally well suppose that the complex metabolic substance of the muscle splits down into simpler substances and that in this transformation energy is generated. Suppose that these simpler substances are poisonous and that they must be removed as rapidly as formed. The rôle of the oxygen may be to oxidise them, thus transforming them into carbon dioxide, an innocuous substance which can be carried away quickly in the blood stream. This line of thought, according to which the rôle of oxygen is an anti-poisonous one, is held at the present day by some physiologists, and many considerations appear to support it; the existence of " oxidases," for instance, enzymes which produce oxidations which would not otherwise occur in their absence. Such enzymes exist in very many tissues, and they may, apparently, be present in an inactive form, requiring the agency of a " kinase " before they are able to act.

The usual view among physiologists is that the muscle fibre is a thermodynamic apparatus transforming the heat generated during metabolism into mechanical energy. How is this transformation effected? It cannot be said that we have any one hypothesis more convincing than another. It has been suggested that alterations of surface tension play a part, or that the heat produced by oxidation causes the fibre to imbibe water and shorten. Engelmann has devised an artificial muscle consisting of a catgut string and an electrical current passing through a coil of wire, and by means of this he has reproduced the phenomena of simple contraction and tetanus. But it remains for future investigation to verify any one of these hypotheses.

When Huxley published his *Physical Basis of Life*, probably few physiologists had any doubt that protoplasm was a definite chemical substance, differing from other organic substances only by its much greater complexity. But in 1880 Reinke and Rodewald published the results of an analysis of the substance of a plant protoplasm and these appear to have demonstrated that the substance was really a mixture of a number of true chemical compounds and was not a single definite one. Now all of these substances might exist apart from protoplasm, and in the lifeless form, and a simple mixture of them could hardly bring forth vital reactions. These results were followed by the morphological study of the cell—the discovery of the architecture of the nucleus, and so on, and so opinion began to turn to the hypothesis that the vital manifestations of protoplasm were the result of its *structure*. Microscopical examination of the cell appeared to disclose a definite arrangement, the " foam " or " froth " of Butschli, for instance. But, again, it was easily shown that the foam, or alveolar structure of protoplasm was merely the expression of physical differences in the substances composing the cell-stuff—they reduced to phenomena of surface tension and the like. Artificial protoplasm and artificial *Amœbæ* were made—at least mixtures of olive oil and various other substances were made which simulated many of the phenomena of protoplasm in much the same way as crystalline products may be made which simulate the growth of a plant stem with its branches. For instance, one has only to shake up a little soapy water in a flask to see what resembles surprisingly the arrangement of certain kinds of connective tissues in the organism. Obviously these artificial phenomena have nothing to do with living substance.

Yet if we grind up a living muscle with some sand in a mortar we do destroy something. The muscle could be made to contract, but after disintegration this power is lost. We have certainly destroyed a structure, or mechanism, of some kind. But, again, the paste of muscle substance and sand still possesses some kind of vital activity, for with certain precautions it can be made to exhibit many of the phenomena of enzyme activity displayed by the intact muscle fibres, or even the entire organism. Mechanical disintegration, therefore, abolishes some of the activities of the organism, but not all of them. If, however, we heat the muscle paste above a certain temperature, the residue of vital phenomena exhibited by it are irreversibly removed, so that heating destroys the mechanism. This we can hardly imagine to be the case (within ordinary limits of temperature at least) with a physical mechanism, but again a mechanism which is partly chemical might be so destroyed. We see, then, that protoplasm possesses a mechanical structure, but that all of its vital activities do not necessarily depend on this structure. The full manifestation of these activities depends on the protoplasmic substance possessing a certain volume or mass, and also on a certain chemical structure.

If living protoplasm has a structure, and is not simply a mixture of chemical compounds, what is it then ? Two or three physico-chemical concepts are at the present time very much in evidence in this connection. When the substances known as *colloids* were fully investigated by the chemists, much attention was paid to them by the physiologists, so that life was called " the chemistry of the colloids," just as after the investigation of the enzymes it was called the " chemistry of the enzymes," and when the discovery of the relative abundance of phosphorus in cell-nuclei and in the brain was discovered, it was called the " chemistry of phosphorus." Colloids (*e.g.* glue) are substances that do not readily diffuse through certain membranes, in opposition to crystalloids (*e.g.* solution of common salt) which do readily so diffuse. They form solutions which easily gelatinise reversibly, that is, can become liquid again (glue) ; or coagulate irreversibly, that is, cannot become liquid again (albumen) ; which have no definite saturation point ; which have a low osmotic pressure (and derived properties), etc. ; and the molecules of which are compound ones consisting of combinations of the molecules of the substance with the molecules of the solvent, or with each other, that is, they are molecular aggregates.

Colloids pass insensibly into crystalloids on the one hand and into coarse suspensions (water shaken up with fine mud, for instance) on the other. We may replace the concept of a colloid by those of " suspensoids " and " emulsoids." A suspensoid is a liquid containing particles in a fine state of division—if the division is that into the separate molecules we have a solution, if into large aggregates of molecules we have a suspension. If the substance in the liquid is itself liquid, the whole is called an emulsoid. On the one hand this approaches to a mixture of oil in soap and water—an emulsion—and on the other hand to such a mixture as chloroform shaken up with water, when the drops of chloroform readily join together so that two layers of liquid (chloroform and water) form.

What we see, then, in protoplasm is a viscid substance possessing a structure of some kind, and containing specialised protoplasmic bodies in its mass (nuclei, nucleoli, granules of various kinds, chlorophyll, and other plastids, etc.). It may contain or exhibit suspensoid or emulsoid parts or substances, or it may contain truly crystalloid solutions. These phases of its constituents are not fixed, but pass into each other during its activity. Nothing that we know about it justifies us in speaking about a " living chemical substance." On analysis we find that it is a mixture of true chemical substances rather than *a* substance. It is no use saying that in order to analyse it we must kill it, for what we can observe in it without destroying its structure or activities indicates that it is chemically heterogeneous.

This is not a textbook of general physiology, and the examples of physico-chemical reactions in the organism which we have selected have been quoted in order to show to what extent the chemical and physical methods applied by the physiologists have succeeded in resolving the activities of the organism. The question for our consideration is this : do these results of physico-chemical analysis fully describe organic functioning ? Dogmatic mechanism says " yes " without equivocation.

Now it is clear, from even the few typical examples that we have quoted, that physiological analysis shows, indeed, a resolution of the activities of the organism into chemical and physical reactions. How could it do otherwise ? How could chemical and physical methods of investigation yield anything else than chemical and physical results ? The fact that these methods *can* be applied to the study of the organism with consistent results shows that their application is valid ; that we are justified in seeing physico-chemical activities in life. But are these results *all* that we have reason to expect ?

We turn now to Bergson's fertile comparison of the physiological analysis of the organism with the action of a cinematograph. If we take a series of photographic snapshots of, *e.g.*, a trotting horse and then superpose these pictures upon each other, we produce all the semblance of the co-ordinated motions of the limbs of the animal. Yet all that is contained in the simulated motion is immobility. From a succession of static conditions we appear to produce a flux. Yet if we could contract our duration of, *e.g.*, a week, into that corresponding to five minutes—if we could speed up our perceptual activity—should we not see the cinematographic pictures as they really are—a series of immovable postures and nothing more : truly an illusion ? If, again, we reverse the direction of motion of the film, we integrate our snapshots into something which is absolutely different from the reality which they at first represented ; and by such devices the illusions and paradoxical effects of the picture-house farces are made possible. Well, then, in the physiological analysis of the activity of the organism do we not do something very analogous to this ? The complexity of even the simplest function of the animal is such that we can only attend to one or two aspects of it at once, arbitrarily neglecting all the rest. We find that the hydrostatic pressure of blood, and lymph, and secretion, the osmotic pressure, the diffusibility, vaso-motor actions, and other things must be investigated when considering the question of how the submaxillary gland secretes saliva. One, or as many as possible, of these reactions are investigated at one time, and then the results are pieced together—integrated—in order to reproduce the full activity of the whole indivisible process. But in doing this do we not introduce something new—a *direction* or order of happening—into the elements of the dissociated activity of the organism ? Each elemental process must occur at just the right time.

What right have we to say that the activity of the organism is *made up* of physico-chemical elements ? Just as much as we have in saying that a curve is made up of infinitesimal straight lines. Let us adopt Bergson's illustration, with a non - essential modification.

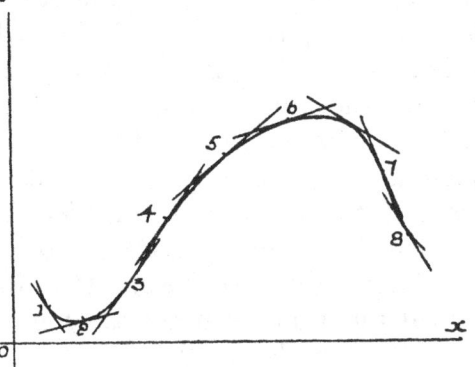

The curve 1-8 is a line which we draw freehand with a single indivisible motion of the hand and arm and eye. It

FIG. 10.

is something unique and individualised, in that no other curve ever drawn, in a similar manner, exactly resembles it. Let us investigate it mathematically. We can select very small portions of it—elements we may call them—and each of these elements, if it is small enough does not differ *sensibly* from a straight line. Let us produce each of these straight lines in both directions, it is then a tangent to the curve, and it does actually coincide with the curve at one mathematical point—the points 1-8 in the figure. The tangent then has *something in common with the curve,* but would a series of infinitesimally small tangents reproduce the curve ?   Obviously not, for the equations of the tangents would have the form $ax + b$, while that of the curve itself would be quite different, containing x as powers of x, or as transcendental functions of x.   In this investigation what we succeed in obtaining are the derivatives of the curve, and to reproduce the latter from its elements we have to integrate the derivatives ; that is, another operation differing in kind from our analytical one must be performed.   Now in this illustration we have doubtless something more than an analogy with our physicochemical analysis of life.   The activities of the organsim do reduce to bio-chemical ones (the elemental straight lines on the curve), and each of these reactions has something in common with life (it is tangent to life, touching it at one point).   But if we attempt to reconstitute life from its physico-chemical derivatives we must integrate the latter, and in doing so we overpass the bounds of physics, just as integrating a mathematical function we necessarily introduce the concept of the " infinitely small."

The physico-chemical reactions into which we dissociate any vital function of the organism have, then, each of them, something in common with the vital function.   But their mere sum is not the function. To reproduce the latter we have to effect a co-ordination and give directions to these reactions.   In all physiological investigations we proceed a certain length with perfect success ; thus the elements, so to speak, of the function of the secretion of saliva are (1) the blood-pressure, (2) the hydrostatic pressure of the secretion in the lumina of the gland tubules, (3) the diffusbility of the substances dissolved in the blood and lymph through the walls of these vessels, (4) the osmotic pressure of the same substances, and (5) the stimulation of the gland cells by "secretory nerve fires." Now the investigations carried out—and no part of the physiology of the mammal has been so patiently studied as the salivary gland—fail, so far, completely to describe the function in terms of these elements. In the end we have to refer the secretion to intra-cellular processes, and then we begin to invoke again processes of osmotic pressure, diffusibility, and so on with reference to the formation of the drops of secretion which we can see formed in the gland cells. We are forced to the formulation of a logical hypothesis as to the nature of these intra-cellular processes, and since much that goes on in the cell substance is, so far, beyond physico-chemical investigation, our hypothesis will be as difficult to disprove as to verify.

Let us return now to Huxley's comparison of the activity of the green plant with the chemical reaction which occurs when an electric spark is passed through a mixture of oxygen and hydrogen. The lecture on the "Physical Basis of Life" was published in 1869; in 1852 William Thomson published his paper "On a Universal Tendency of Nature to Dissipation of Energy," and a year or two before that Clausius had applied Carnot's law to the kinetic theory of heat : the second principle of energetics had therefore even then been exactly formulated, but its significance for biological speculation had not been recognised by Huxley, any more than it has generally been recognised by most biologists since 1869. What, then, does the comparison of Huxley show ? Clearly that the physical changes which occur in the explosion of a mixture of oxygen and hydrogen *trend* in a different direction from those which occur in the photo-synthesis

H

of starch by a green plant. Generally speaking, chemical activity, that is, the possibility of occurrence of chemical reactions, is a case of the second law of energetics. Energy passes from a state of high to a state of low potential. A chemical reaction will occur if this change of potential is possible.

In all such changes energy is dissipated. What exactly does this mean? It means that, generally speaking, the potential energy of chemical compounds tends to transform into kinetic energy; while differences in the intensity factor of the kinetic energy of the bodies forming a system tend to become minimal. In a mixture of oxygen and hydrogen there is energy of two kinds, (1) potential energy due to the position of the molecules (O and H molecules are separated); and (2) kinetic energy of the molecules (which are moving about in the masses of gas). After the explosion the potential energy acquired in the separation of the molecules of O and H has disappeared (the molecules having combined to form water), but the kinetic energy has greatly increased, since the explosion results in the formation of steam at high temperature. But now this steam radiates off heat to adjacent bodies, or becomes cooled by direct contact with the envelope which contains it. The energy of the explosion is therefore distributed to the adjoining bodies, and the temperature of the latter becomes raised. But these again radiate and conduct heat to other bodies, and in this way the heat generated becomes indefinitely diffused.

The general effect of all physico-chemical changes is therefore the generation of heat, and then this heat tends to distribute itself throughout the whole system of bodies in which the physico-chemical changes occur. The energy passes into the state of kinetic energy, that is, the motion of the molecules of the bodies to which the heat is communicated. This molecular motion is least in solids, greater in liquids, and greatest in gases. If solids, liquids, and gases are in contact, forming complex systems, the kinetic energy of their molecules becomes distributed in definite ways, depending on the constants of the systems. After this redistribution the kinetic energy of these molecules is unavailable for further energy transformations, so that phenomena or change in the system ceases. There is no longer effective physical diversity among the parts of the system.

We find that this conception of dissipation of energy cannot be applied to the organism, at least not with the generality in which it applies to physical systems. Why? Not because the conception is unsound, or because the physico-chemical reactions that occur in material of the organism are of a different order from those that occur in inorganic systems—they are of the same order. The second law of energetics is subject to limitations, and it is because it is applied to organic happenings without regard to these limitations that it does not describe the activities of the organism as well as it describes those of inorganic nature.

What, then, are these limitations? We note in the first place that the laws of thermodynamics apply to bodies of a certain range of size ; or at least the possibility of mathematical investigation (on which, of course, all depends) is limited to " differential elements " of mass, energy, and time. We cannot apply mathematical analysis to bodies, or time-intervals of " finite size," since the methods of the differential and integral calculus would not strictly be applicable. But molecules are so small (1 cubic centimetre of a gas may contain about 5.4×10^{19} of them) that even such a minute part of a body, or liquid, or gas as approximates to the infinitesimally small dimensions required by the calculus, contains an enormous number of molecules.

Obviously we cannot investigate the individual molecules. Even if experimental methods could be so applied, such concepts as density, pressure, volume, or temperature would have no meaning. Physics, then, is based on collections of molecules, and the properties of a body are not those of a molecule of the same body. Such concepts as temperature and pressure are *statistical* ones, and are applied to the mean properties of a large number of molecules.

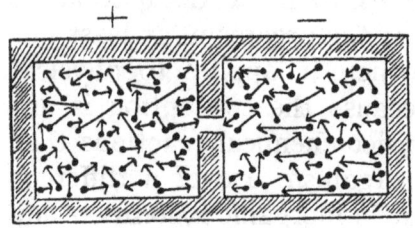

FIG. 11.

We can best illustrate this by considering Maxwell's famous fiction of the "sorting demons." Let us imagine a mass of gas contained in a vessel the walls of which do not conduct heat. Let there be a partition in this vessel also of non-conducting material, and let there be an aperture in this partition greater in area than a molecule, but smaller than the mean free path of a molecule. Now this mass of gas has a certain temperature which is proportional to the mean velocity of movement of the molecules. The second law says that heat cannot pass from a cold region in a system to a hot region without work being done on the system from outside, nor can an inequality of temperature be produced in a mass of gas or liquid except under a similar condition. But " conceive a being," says Maxwell, " whose faculties are so sharpened that he can follow every molecule in its course ; such a being, whose attributes are still as essentially finite as our own, would be able to do what is at present impossible to us." [1] For the temperature of the gas depends on the velocities of the molecules, and in any part of the gas these velocities are very different. Suppose that the demon saw a molecule approach which was moving at a much greater velocity than the mean : he would then open the door in the aperture and let it pass through from - to +. On the other hand, should a molecule moving at a velocity much less than the mean approach he would let it pass from + to - . In this way he would sort out molecules of high from those of low velocity. But the collisions between the molecules in either division of the vessel would continually produce diversity of individual velocity, and in this way the difference of temperature between + and - would continually be increased. Heat would thus flow from a region of low to a region of high temperature without an equivalent amount of work being expended.

Now we must not introduce demonology into science, so, lest this fiction of Maxwell's should savour of mysticism, or something equally repugnant, we shall state the idea involved in it in quite unexceptionable terms. The conclusions of physics are founded on the assumption that we cannot control the motions of individual molecules. In a mass of gas, or liquid, or in a solid, the molecules are free to move and do move. Their individual velocities and free paths vary considerably from each other. These motions and paths are un-co-ordinated—" helter-skelter "—if we

---

[1] Impossible, in the sense that while we are unable to " abrogate " a physical law, Maxwell's finite demon could, although his faculties were similar in nature to ours.

like so to term them. Physics considers only the statistical *mean* velocities and free paths. The irreversibility of physical phenomena, the fact that energy tends to dissipate itself, the second law of thermodynamics, depend on the assumption that Maxwell's demons exist only in imagination. We must appeal to experience now. There is no *a priori* reason why the phenomena of physics should be directed one way and not the other, for it is possible to conceive a condition of our Universe in which, for instance, solid iron would fuse when exposed to the atmosphere. In such conditions organisms would grow backwards from old age to birth, with conscious knowledge of the future but no recollections of the past. Experience shows, however, that phenomena do tend in one way—*but this experience is that of experimental physics*, so that for the latter science Maxwell's demons do not exist. Now physiology has borrowed from physics, not only the experimental methods, but also the fundamental concepts of thermodynamics. The organism, therefore (so physiology must conclude), cannot control the motions of individual molecules, and so vital processes are irreversible. But we have seen that the processes of terrestrial life as a whole are reversible, or tend to reversibility. We must therefore seek for evidence that the organism *can* control the, otherwise, un-co-ordinated motions of the individual molecules.

The Brownian movement of very small particles of matter is so familiar to the biologist that we need not describe it. It is doubtless due to the impact of the molecules of the liquid in which the particles are suspended. Groups of molecules travelling at velocities above the mean hit the particle now on one side, and again on the other, and so produce the peculiar trembling which Brown thought was life. Now the particle must be below a certain size in order to be so affected. Are there organisms of this size? Undoubtedly there are, for many bacilli show Brownian movements, while we have reasons for believing that ultra-microscopic organisms exist. Also, on the mechanistic hypothesis there are " biophors," the size of which is of the same order as that of the molecules of the more complex organic compounds. All these must be affected by the molecular impacts of the liquid in which they are suspended. Can they distinguish between the impacts of high-velocity molecules and those of mean-velocity ones, and can they utilise the surplus energy of the former ? This has been suggested by the physicists. In Brownian movement, says Poincaré, "we can almost see Maxwell's demons at work."

The suggestion is not merely a speculative one, for it is well within the region of experiment. To prove it experimentally we should only have to show that the temperature of a heat-insulated culture of prototrophic bacteria falls while the organisms multiply.

Is it not strange that the biologists, to whom the Brownian movement is so familiar, should have failed to see its possibly enormous significance ? Is it not strange that the biologists, to whom the distinction between the statistical and individual methods of investigation is so familiar, should have failed to appreciate this distinction when it was made by the physicists ? Is it not strange that while we see that most of our human effort is that of *directing* natural agencies and energies into paths which they would not otherwise take, we should yet have failed to think of primitive organisms, or even of the tissue elements in the bodies of the higher organisms, as possessing also this power of directing physico-chemical processes ?

CHAPTER IV

Two main conclusions emerge from the discussions of the last three chapters : (1) that physiology encourages no notions as to a " vital principle " or force, or form of energy peculiar to the organism ; and (2) that although physiological analysis resolves the metabolism of the plant and animal body into physico-chemical reactions, yet the direction taken by these is not that taken by corresponding reactions occurring in inorganic materials. From these two main conclusions we have, therefore, to construct a conception of the organism which shall be other than that of a physico-chemical mechanism.

The ordinary person, unacquainted with the results of physiological analysis, and knowing only the general modes of functioning of the human organism, has, probably, no doubt at all that it is " animated " by a principle or agency which has no counterpart in the inorganic world. This is the " natural " conclusion, and the other one, that life is only an affair of physics and chemistry, must appear altogether fanciful to anyone who knows no more than that the heart propels the blood, that the latter is " purified " in the lungs, that the stomach and liver secrete substances which digest the food, and so on. It is difficult for the modern student of biology, saturated with notions of biochemical activities, gels and sols and colloids and reversible enzymes and kinases and the like, to realise that the belief in a vital agency is an intuitive one, and that the mechanistic conception of life is only the result of the extension to biology of *methods* of investigation, and not a legitimate conclusion from their *results*.

To the anatomist, the embryologist, and the naturalist, as well as to the physicist unacquainted with the details of physiology, no less than to the ordinary person this is perhaps by far the most general attitude of mind. It would probably be impossible for anyone to study only organic form and habits and come to any other conclusion than that there was something immanent in the organism entirely different from the agencies which, for instance, shape continents, or deltas, or river valleys. And this conclusion would probably come with still greater force to the embryologist, even though he still possessed a general knowledge of physiological science.

The mechanistic conception of life has, without doubt, been the result of the success of a method of analysis. One sees clearly that just in proportion as physical and chemical sciences have been most prolific of discovery, so physiology, leaning upon them and borrowing their methods, has been most progressive and mechanistic.

Mechanistic hypotheses of the organism may all be traced back to Descartes, who built upon the work of Galileo and Harvey. The anatomy of Vesalius and his successors would have led to no such notions, had not the discoveries of Copernicus, Tycho, and Kepler shown men an universe actuated by mechanical law. To a thinker like Descartes, at once the very type of philosopher and man of science, Harvey's discovery of the circulation of the blood must have suggested irresistibly the extension of mechanical law to the functioning of the human organism, and it is significant that he made this extension without including a single chemical idea, and yet produced a logical hypothesis of life as satisfactory and complete in its day as, for instance, the Weismannian hypothesis of heredity has been in ours.

His hypothesis of the organism was purely mechanical. It has been said that his organism was an automaton, like the mechanical Diana of the palace gardens which hid among the rose-bushes when the foot of a prying stranger pressed upon the springs hidden in the ground. Its functions were matters of hydraulics: of heat, and fluids, and valves. His physiology was Galenic, apart from Harvey's discovery of the motion of the blood in a circuit, for he did not accept the notion of the heart as a propulsive apparatus. The food of the intestine was absorbed as chyle by the blood and carried to the liver, where it became endued with the " natural spirits," and then passing to the heart it became charged with the " vital spirits " by virtue of the flame, or innate heat, of the heart, and the action of the lungs. This flame of the heart, fed by the natural spirits, expanded and rarefied the blood, and the expansion of the fluid produced a motion, which, directed by the valves of the heart and great vessels, became the circulation. The more rarefied parts of the blood ascended to the brain, and there, in the ventricles, became the " animal spirits."

Subtle and rarefied though they were, these animal spirits were a fluid, amenable to all the laws of hydrodynamics. This was contained in the cerebral ventricles, and its flow was regulated just like the water in the pipes and fountains of the garden mechanisms. From the brain it flowed through the nerves, which

were delicate tubes in communication with the
ventricles, and which were provided with valves ; and
this outward flow corresponds to our modern efferent
nervous impulse. The afferent impulse was represented
by the action of the axial threads contained in the
nerve tubuli. When a sensory surface was stimulated,
these threads became pulled, and the pull, acting on
the wall of the cerebral ventricle, caused a valve to
open and allowed animal spirits to flow along the
nerve to all the parts of the body supplied by the
latter. In the effector organs, muscles or glands, this
influx of animal spirits produced motion or other
effects. This, in brief, was the physiology of Descartes.

He spoiled it, says Huxley, by his conception of
the " rational soul." Fearing the fate of Galileo, he
introduced the soul into his philosophy of the organism
as a sop to the Cerberus of the Church. It was un-
worthy : a sacrifice of the truth which he saw clearly.
Is it likely that Descartes deliberately made part of
his philosophy antagonistic to the rest with the object
of averting the censure of the Church ? He was not
a man likely to rush upon disaster, but the conviction
that what he wrote had in it something great and
lasting must have made it hardly possible that he
should traffic with what he held to be the truth.

The rational soul was something superadded to
the bodily mechanism. It was not a part of the body
though it was placed in the pineal gland ; a part of
the brain, which by its sequestered situation and rich
blood supply suggested itself as the seat of some
important and mysterious function. Its existence
was bound up with the integrity of the body, and on
the death of the latter the soul departed. But the body
did not die because the soul quitted it, it had rather
become an unfit habitation for the soul. Without

the latter the functions of the healthy body might still proceed automatically, and if the soul influenced action it actuated an existing mechanism, and without that mechanism it could not act, though the mechanism might act without the soul. Thought, understanding, feeling, will, imagination, memory, these were the prerogatives of the soul, and not those of the automatic body. But the movements of the latter, even voluntary movements, depended on a proper disposition of organs, and without this they were wanting or imperfect.

Thus to a thoroughgoing mechanism Descartes joined a spiritualistic and immortal entity ; and this, to the materialism of the middle of the nineteenth century, was the blemish on his philosophy. Now of all men who have ever lived he is probably the one who has most profoundly influenced modern thought and investigation : to us what he wrote seems strangely modern, and this apparently arbitrary association of spiritualistic and materialistic elements in life seems almost the most modern thing in his writings. Being, he said, was indeed thought, but how could he derive thought from his clockwork body, with its valves and conduits and wires ? No more can we derive consciousness from the wave of molecular disturbance passing through afferent nerve and cerebral tracts. We must account for all the energy of this disturbance, from its origin in the receptor organ to its transformation into the wave of chemical reaction in the muscle, and we must regard its transmission as a conservative process. But how does the state of consciousness accompanying the passage through the cortex of this molecular disturbance come into existence ? None of the energy of the nerve disturbance has been transformed into consciousness : the latter is not energy nor anything physical. It is something concomitant with the physico-chemical events involved in a nervous process, an " epiphenomenon." We have to imagine a " parallelism " between the mechanistic body and the mind. But if we admit that consciousness may be an effective agency in our behaviour, what is the difference between modern theories of physico-psychic parallelism and the Cartesian theory of a rational soul in association with an automatic body ? Descartes denied the existence in animals other than man of the rational soul ; the latter was not necessary. But he, like us, must have been familiar with reflex actions and must have seen that consciousness was not invariably associated, even in himself, with bodily activity. And he must have recognised the great distinction between the intelligent acting of man and the instinctive behaviour of the lower animals. There was something in man that was not in the brute.

Thus the first physiology, borrowing its ideas and methods from the first physics, was, like the latter, a mechanical science. After Galileo and Torricelli came Borelli with his purely mechanical conceptions of animal movement, and of the blood circulation, introducing even then mathematics into biology. There was no chemistry in these speculations, though Basil Valentine and Paracelsus and Van Helmont had preceded Descartes and Borelli. This chemistry was mystical, and though chemical reactions had been studied in the organism, they were supposed to be controlled by spiritual agencies, the " archei " of the first bio-chemists. But that notion was to disappear, and with Sylvius the conception of the animal body as a chemical mechanism arose. All that was valuable in Van Helmont's chemistry was taken up by Sylvius, but in his mind the fermentations of the older chemists were sufficient in themselves without the mystical " sensitive soul " and " archei." With Sylvius and Mayow physiology became based upon chemical discovery and again became mechanistic, and remained so until the time of Stahl, when chemical discovery attained for the time its greatest development.

The seventeenth century ended with the work of Stahl. It is well known to students of science how the views of this great chemist sterilised chemical investigation almost until the time of Lavoisier. The notion of phlogiston as an active constituent of material bodies entering and leaving them in their reactions with each other was a clear and simple one, and it served as a working hypothesis for the chemists who immediately followed Stahl. It was, of course, a false hypothesis, and retarded discovery to the extent that the greater part of the eighteenth century is a blank for chemistry, when compared with the seventeenth and nineteenth centuries. Deprived therefore of the stimulus afforded by new physico-chemical methods of investigation, physiology ceased to maintain the progress it had made during the previous century, and the only great name of this period is that of von Haller. Comparative anatomy, and zoological exploration, on the other hand, made enormous advances, and for these branches of biology the eighteenth century was the great period. It was the period of the historic vitalistic views—vital principles, and vital and formative forces. Stahl's teaching dominated physiology just as it did chemistry. Chemical and physical reactions occurred in the living body just as they did in non-living matter, but they were controlled and modified by the soul, or vital principle. It has been said that Stahl's vitalistic teaching retarded the progress of physiology, but it does not seem clear that this was the case. What did retard physiological discovery was the lack of progress made by chemistry and physics, and this may have been the result of the Stahlian phlogistic hypothesis.

However this may be, it seems clear that it was the discoveries of the great chemists of the close of the eighteenth century that again introduced mechanistic views into physiology. With the discoveries of Lavoisier and his successors the latter science acquired new methods of research and the older working hypotheses were re-introduced. There has been no recession from this position during the nineteenth century. Mechanistic biology culminated in the writings of Huxley and Max Verworn and received a new accession of strength almost in our own day in the modern discoveries of physical chemistry; and when physiology became truly a comparative science, and embraced the lower invertebrates, it became perhaps most mechanistic—witness the writings of Jacques Loeb.

Of far greater philosophical importance than the physico-chemical investigation of the functioning of individual organisms has been the essentially modern experimental study of embryological processes. The former deals essentially with the *means* of growth, reproduction, and so on. We can no longer doubt that the changes which we can observe taking place in the organism, either the developing embryo or the fully formed animal, are, in the long run, physico-chemical changes; and in ultimate analysis we cannot expect to find anything else than processes of this nature.

But physiological investigation has failed to provide anything more than this analysis. If we watch the development of the egg of an animal into a larval form, and continue to trace the metamorphosis of the larva into the perfect animal, we cannot fail to conclude that, beside the individual physico-chemical reactions which proceed, there is also *organisation*. The elementary processes must be *integrated*. There must be a due order and succession in them. In studying developmental processes, in considering the developing organism *as a whole*, we are impressed above all else with the notion that not only do physico-chemical reactions occur, but that these are *marshalled* into place, so to speak. When we attempt to make a description of this integration of those ultimate processes which we can describe in terms of physical chemistry, physiology fails us. " At present," says Morgan, " we cannot see how any known principles of chemistry or of physics can explain the development of a definite *form* by the organism or by a piece of the organism." It is true that we can attempt to imagine a physico-chemical mechanism which is the organisation of the developing embryo ; but this must be a logically constructed mechanism, not only incapable of experimental verification, but which can also be demonstrated, purely by physical arguments, to be false. This conclusion may, without exaggeration, be said to be that of modern experimental embryology.

There have always been (in modern times) two views as to the nature of the embryological process : (1) that the egg contained the fully formed organism in a kind of rolled-up condition, and that the process of development consisted merely in the unfolding (*evolution*) of this embryonic organism, and in the increase in volume of its parts. This was the hypothesis of preformation held in the beginning of embryological science. It involved various consequences : the limitation, for instance, of the duration of a species, since each generation of female organisms contained in their

ovaries all the future generations ; with other consequences which the preformationists did not hesitate
to accept. (2) The other view was the later one
of epigenesis : the egg was truly homogeneous and
the embryo grew from it. Obviously the acceptance
of this hypothesis led to vitalism, and we find that it
was abandoned just as soon as the embryologists recognised that physics provided a corpuscular theory of
matter, when a return was made to the preformation
views of earlier times ; views which lent themselves to
the construction of a
mechanistic hypothesis
of development.

We may state very
briefly the main facts of
the development of a
typical animal ovum,
such as that of the sea-
urchin.

The fertilised ovum
divides into two (2), and
then each of these blastomeres divides again in a plane
perpendicular to the first division plane (3). The third
division plane is at right angles to the first two, and it
cuts off a tier of smaller blastomeres from the tops of the
first four. There are now (4) two tiers of blastomeres,
a lower tier of large blastomeres and an upper tier
of smaller ones. This is the 8-cell stage. Next,
each of these blastomeres divides in two simultaneously
so that the embryo now consists of sixteen cells. After
this the divisions proceed with less regularity, but
after about ten divisions the embryo consists of about
1000 cells (2^{10}), and these are arranged to form a
hollow sphere consisting of a single layer of cells. The
latter are furnished with cilia, and the whole embryo,

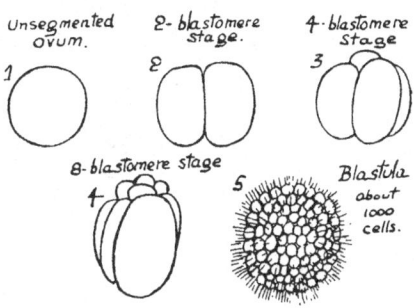

FIG. 12.

I

now known as the blastula, can swim about by the movements of these cilia. Further development results in another larval form—the gastrula, and yet another, the pluteus larva. After this the transformation into the fully formed sea-urchin occurs.

With various modifications this scheme represents the early development of a very large number of animals belonging to most groups.

If we study the process of cell-division we shall find it very complicated. The ovum, immediately after fertilisation, consists of two main parts, the nucleus and the cytoplasm.

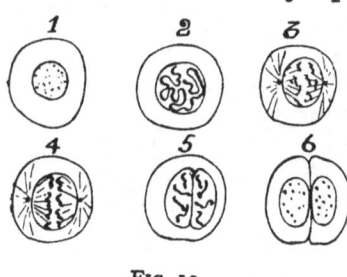

FIG. 13.

Within the nucleus is a substance distinguishable from the rest; it is distributed in granules and is called the chromatin (1). When the cell is about to divide this chromatin becomes arranged in a long coiled thread (2), and then (3) this chromatic thread breaks into short rods called chromosomes. Two little granules now appear, one at each end of the nucleus, and very delicate threads, the asters, appear to pass from each of these bodies towards the chromosomes (4). Each of the latter then splits lengthways into two, and a half chromosome appears to be drawn by the asters towards the poles of the nucleus. The latter then divides (5) and then the whole cell divides. What thus, in essence, happens in nuclear divisions is that the chromatin of the nucleus is more or less accurately halved. Apparently this substance consists of very minute granules and the whole process is directed towards the splitting of each of these granules into two. A half-granule then goes to each of the daughter nuclei.

Every time the embryo divides this process is repeated. Thus each of the (theoretically) 1028 cells of the blastula contains $\frac{1}{1028}$th of the substance of each chromatic granule in the fertilised ovum.

Pfluger and Roux (in 1883 and 1888 respectively) were the pioneers in the experimental study of the development of the ovum, and the results of their work and that of their successors has, more than anything else in biology, modified and shaped our notions of the activities of the organism. Roux found, or thought so at least, that the first division of the frog's egg marked out the right and left halves of the body, the one blastomere giving rise to the right half, the other to the left half. The next division, which separates each of these blastomeres, marked out the anterior and posterior parts of the embryo. Thus :—

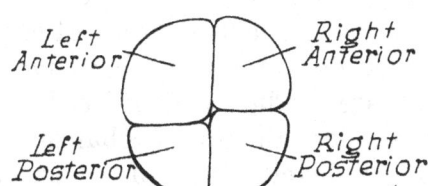

Fig. 14.—The frog's egg in the 4-blastomere stage seen from the top.

Now in an experiment which has become classical Roux succeeded in killing one of the blastomeres in the 2-cell stage, while the other remained alive. The uninjured blastomere then continued to develop, *but it gave rise to a half-embryo only.*

Upon these experiments the Roux-Weismann hypothesis of development—the " Mosaik-Theorie "—was developed. The lay reader will see how obviously the facts of nuclear division and the experimental results indicated above lend themselves to a mechanistic hypothesis. Notice that but for the physical conception of matter as made up of molecules and atoms the mosaic-theory would hardly have shaped itself in the minds of biologists. But this notion of matter consisting of corpuscles must have suggested that the essential " living material " of the organism consisted also of corpuscles, as soon as a microscope powerful enough to see the chromatic granules was turned on a dividing cell prepared so as to render these bodies visible. Obviously the primordial ovum contained all the elements of the organisms into which it was going to develop. But then in the process of division of the ovum all these chromatic granules are shared out among the cells, and a really very pretty mechanism comes into existence for this purpose of distribution.

Weismann built up his hypothesis of the germ-plasm upon the observations we have outlined. The chromatic matter of the nucleus consists of elements called *determinants*, the determinants themselves being composed of ultimate bodies called *biophors*. Each determinant possesses all the mechanism, or factors, necessary for the development of a part of the body : there are determinants for muscles, nerves, connective tissues, for the retina of the eye, for hairs of each colour, for the nails, and so on. All these determinants are contained in the chromatin of the nucleus of the egg, and in the divisions of the latter they are gradually separated so that ultimately each cell of the larva contains the determinants for one individual part, or organ, or organ-system of the adult body. The right blastomere, for instance, contains all the determinants for the right side of the frog's body, those for the left side being contained in the left half. The process of cell-division involved in the segmentation of the egg consists then in the orderly disintegration of this complex of determinants, and in the marshalling into place of the isolated elements. The cell body—the cytoplasm—carried out a very subordinate rôle, mainly that of nourishing the essential chromatic substance. Such was the Roux-Weismann Mosaic-theory of development in its pristine form.

It is clearly a preformation hypothesis. It is true that the actual organism is not contained in the germ, but all the parts of the latter, even the colours of the eyes or hair, are present in it in the form of the determinants. Obviously it involves a mechanism of almost incredible complexity. But if we regard it as a working hypothesis of development this complexity of detail does not matter ; its truth would be indicated by the fact that all analysis of the processes involved would tend to simplify it and to smooth out the complexity. But this is exactly what has not happened, for all subsequent investigation has necessitated subsidiary hypothesis after hypothesis. As a theory of development it has failed entirely.

If, after one of the blastomeres in the frog's egg at the 2-cell stage be killed, the egg is then turned upside down, the results of the experiment become totally different ; the uninjured blastomere develops into a *whole* embryo, differing from the normal one chiefly in that it is smaller. If the uninjured egg in the 2-cell stage be turned upside down *two whole embryos*, connected together in various ways, develop. In the frog's egg the two first blastomeres cannot be separated from each other without rupturing them, but in the egg of the salamander they can be separated. After this separation two perfect, but small, embryos develop. In the egg of the newt a fine thread can be tied round the furrow formed by the first division. If this ligature be tied loosely it does not affect development, and then it can be seen that the median longitudinal plane of the embryo does not correspond, except by chance, with the first division plane. If the ligature be tied tightly, then each of the blastomeres gives rise to an entire embryo. If it is tied in various places monsters of various types are produced. Therefore there is no segregation of the determinants in the first two blastomeres. These results, moreover, are not exceptional, for similar ones have been obtained with other animal embryos, in fishes, *Amphioxus*, ascidians, medusæ, and hydrozoa, and in some cases even each of the first four blastomeres develops into an entire embryo when it is separated from the rest. In the sea-urchin embryo the blastomeres can be shaken apart ; or by removing the calcium which is contained in sea water the blastomeres can easily be separated from each other. It was then found by Driesch that each of the blastomeres in the 16-cell stage could develop into an entire embryo. It is plain, then, that up to this stage at least there has been no segregation of the determinants.

Upon the results of these experiments Driesch based his first proof of vitalism. Let us suppose that there is a mechanism in the developing egg. Now the embryo which results from the latter sooner or later acquires a three-dimensional arrangement of parts : head-end differs from tail-end, dorsal surface differs from ventral surface, and the parts differ on either side of the median plane. The mechanism must, therefore, be one which acts in three dimensions, anterior and posterior, laterally, and dorso-ventrally. We may represent it by a diagram of three co-ordinate axes, x, y, z ; x and y being in the plane of the paper, and z at right angles to the plane of the paper. Now in the 2-cell stage the same mechanism must be present, for this stage develops normally into one entire embryo.

But since *either* of the blastomeres may develop into an entire embryo, the mechanism must also be present in each of them, and since in the 16-cell stage each blastomere may develop an entire embryo, it must be present in each of the sixteen blastomeres. A three-dimensional mechanism is therefore capable of division down to certain limits.

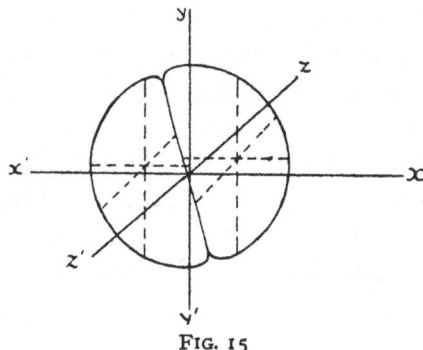

FIG. 15

Suppose now that we allow the sea-urchin egg to develop normally up to the blastula stage. In this stage it is a hollow sphere, the wall of which is a single layer of cells. It is similar all round, that is, we cannot distinguish between top and bottom, right and left, anterior and posterior regions; but since it develops into a larva in which all these distinctions become apparent very soon, it must possess the three-dimensional mechanism, since the activity of the developmental process is going to produce different structures in each direction.

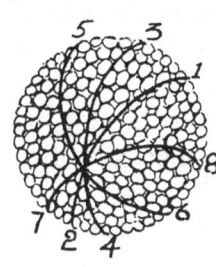

FIG. 16.

Now the blastula, by very careful manipulation can be divided, cut into parts with a sharp knife. Since it is similar all round the direction of the cut is purely a matter of chance. It can be cut through along the planes 1 2, 3 4, 5 6, 7 8, for instance; really there are an infinite number of planes along which the blastula can be cut into two separate parts, and the direction of the plane is not a matter of choice, but purely a matter of chance. Nevertheless, each of the parts into which the larva is cut becomes an entire embryo. For a time the partial blastula—approximately a hollow hemisphere in form—goes on developing as if it were going to become a partial embryo, but soon the opening closes up and development becomes normal. It does not matter even if the two parts into which it is divided are not alike in size ; provided that a part is not too small, it will follow the ordinary course of development.

Suppose the blastula opened out on the flat, like the Mercator projection of a globe on a flat map. Suppose that a is a small element of it. Suppose that the rectangles $bcde, FGHe, IJcL, MNoe$, and as many more as we care to make, represent the pieces of the blastular wall separated by our operation

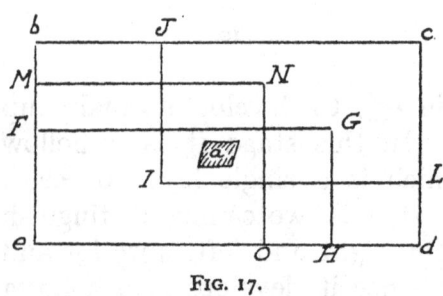

FIG. 17.

—they all contain the element a, but this is in a different position in each case. There are really an infinite number of such parts of the blastula and a occupies an infinitely variable position in each of them.

This demonstration is very important, so let us make it as clear as possible :   Driesch's logical proof of vitalism may be stated as follows :—

The different parts of the blastula are going to become different parts of an embryo.

The part a, occupying a definite position in the entire blastula, is going to become a definite part, having a definite position, in the embryo ;

But each partial blastula becomes an entire embryo and the same part *a* occupies a different position in each.

Therefore *any* part of the blastula may become *any* part of the embryo.

Now if a mechanism is involved, it must, according to our ideas of mechanism, be one which is different in its parts, for each part of it produces a different result from the others ;

But since any part of the mechanism may produce any of the different results contained in the embryo, every one of its parts must be similar to every other one.

That is, all the parts of the mechanism are the same, though the hypothesis requires that they should be different.

We conclude, then, that a mechanism such as we understand a mechanism to be in the physical sciences cannot be present in the developing ovum.

Nevertheless, an *organisation*, using this term as an ill-defined one for the present, must exist in the ovum, or the system of undifferentiated cells into which the ovum divides, during the first stages of segmentation. In certain animals, Ctenophores (Chun, Driesch, and Morgan), and Mollusca (Crampton), for instance, separation of the blastomeres in the first stages of segmentation produces different results from those mentioned above. In these cases the isolated blastomeres develop as partial embryos, that is, the latter are incomplete in certain respects, and this incompleteness corresponds, in a general way, to the incompleteness of the part of the ovum undergoing development. We have thus the apparently contradictory results : (1) each of the first few blastomeres resulting from the first divisions of the ovum is similar to the entire ovum, and develops like it ; and (2) each of the first few blastomeres is different from the others, and from the entire ovum, and develops differently from the others, and from the entire ovum.

Let us try to construct a notion of what this organisation in the developing ovum must be. In the 16-blastomere stage of the sea-urchin egg we have a " system " of parts. In the case of normal development each of these parts has a certain actual fate— it will form a part of the larva into which the embryo is going to develop : It has, as Driesch says, a *prospective value*. But let the normal process be interfered with, and then each of these parts does something else. In the extreme case of interference, when the blastomeres are separated from each other, each blastomere, instead of forming only a part of a larva, forms a whole larva. The *prospective potency* of the part, that is its possible fate, is greater than its prospective value. Normally it has a limited, definite function in development, but if necessary it may greatly exceed this function.

What any one blastomere in the system will become depends upon its position with regard to the other blastomeres. When the egg of the frog is floating freely in water it lies in a certain position with the lighter part uppermost, and then development is normal, each of the two first blastomeres giving rise to a particular part of the body of the larva ; that is, each of them is affected by the contact of the other and develops into whatever part of the normal embryo the other does not. But let the egg in the 2-cell stage be turned over and held so that the heavy part is uppermost : the protoplasm then begins to rotate so as to bring the lighter part uppermost; but the two blastomeres do not, as a rule, adjust themselves to the same extent, and at the same rate, and corresponding parts may fail to come into contact with each other. Lacking, then, the normal stimulus of the other part, each blastomere begins to develop by itself, and a double embryo is produced. It is clear, then, both from this case and the last one, that the actual fate of any one part of the system of blastomeres *is a function of its position.* What it will become depends precisely on where it is situated with respect to the other parts.

Driesch, then, calls the system of parts in such cases as the 2-cell frog embryo, or the 16-cell sea-urchin embryo, an *equipotential system,* since each part is potentially able to do what any other part may do, and what the whole system may do. But in normal development each part has a definite fate and its activity is co-ordinated with that of all the other parts. It is, therefore, an *harmonious equipotential system,* each part acting in harmony, and towards a definite result, with all the others ; although if necessary it can take the place of *any* or *all* of the others.

Such an harmonious equipotential system exists only at the beginning of the development of the egg. It is represented by the 8-cell stage of Echinus but not by the 16-cell stage, since, though the $\frac{1}{16}$-blastomeres produce gastrulæ (the first larval stage), they do not produce plutei (the second stage). It is represented by the 4-cell stage of Amphioxus but not by the 8-cell stage. It is not exhibited even by the 2-cell stage of the Ctenophore egg. What does this mean ? It means that the further development proceeds, the less complete does the " organisation " inherent in any one part of the system become. " The ontogeny assumes more and more the character of a mosaic work as it proceeds " (Wilson).

Or perhaps it means, and this is the better way of putting it, that the " organisation," whatever it may be, depends on size. We see this very clearly in the experiment of cutting in two the blastula of the sea-urchin. If the pieces are of approximately equal size each will form an entire Pluteus larva, but if one of them is below a certain limit of size it will not continue to develop. The " organisation," therefore, has a certain volume, and this volume is much greater than that of any one of the cells of which the fragment exhibiting it is composed. It is enormously greater than the volume of any group of determinants which we can imagine to represent the different kinds of cells composing the body of the Pluteus larva, and still more enormously greater than the volume of a " molecule " of protoplasm. Now this association of " organisation " and size is of immense philosophical importance, for it does away, once and for all, with the idea that the " organisation " is solely a series of chemical reactions. If it were, one cell of the blastula would contain it, for on the mechanistic hypothesis one cell, the egg-cell, contains it, and this cell can be divided innumerable times and still contain it. The egg is a *complex equipotential system* (Driesch), which divides again and again throughout innumerable generations, and still contains the " organisation."

It is in vain that we attempt the misleading analogy of the " mass action " of physical chemistry, to show that volume may influence chemical action. In such a mass action what we have is this :—

$$A_a + B_b \rightleftharpoons C_c + D_d$$

the letters A, B and C standing for chemical substances present, and the letters a and b, etc., representing the active masses of these substances. But variations in this active mass affect only the *velocity* of the reaction. What we have to account for in our blastula experiments is the *nature* of the reaction, and how can velocity or even nature of reaction affect *form* ? If we could show that the form of the crystals deposited from a solution in some reaction depended on the volume of the solution, the analogy would be closer, though even then the difficulties in pressing it would be so enormous as to render it futile to attempt to entertain it.

A chemical mechanism cannot, then, be imagined, much less described, and the only other mechanism so far suggested is the Roux-Weismann one, involving the disintegration of the determinants supposed to be present in the egg nucleus. Let us suppose (in spite of the incredible difficulty in so doing) that there is such a mechanism. It must usher the nuclei containing the determinants of the embryonic structure into their places: those for the formation of the nerve-centre go forward ; those for the mouth, gut, and anus go backwards and downwards ; those for the arms go forwards, ventrally, and posteriorly, in a very definite way ; and those for the complicated skeleton are distributed in a variety of directions which defy description. These nuclei are, in short, moved up and down, right and left, backwards and forwards, and become built up into a complicated architecture. Suppose we prevent this. Suppose we compress the segmenting egg between glass plates so that the nuclei are compelled to distribute themselves in one plane only : to form a flattened disc in which the only directions are right and left and anterior and posterior. This has been done by Driesch and others. On the Roux-Weismann original hypothesis a monstrous larva ought to result, for the first nuclei separated from each other have been forced into positions altogether different from those which they should have occupied had they developed normally. Yet on releasing the pressure readjustment takes place. New divisions occur so as to restore the normal form of larva. The Roux-Weismann subsidiary hypothesis is that the stimulus of the pressure has compelled the nuclei to divide at first in such a way as to compensate for the disturbance.

Let us remove some of the blastomeres. On the original hypothesis the determinants for the structures which the nuclei of these blastomeres contained have been lost. These structures should, therefore, be missing in the embryo. But nothing of the sort is the result. Other nuclei divide and replace the lost ones, and the embryo develops as in the normal mode. The reply is that in addition to the determinants which were necessary for their own peculiar function, these nuclei contained a reserve of all others. On disturbance these determinants, " latent " in all other conditions, became active and restituted the lost parts.

Let us remove some organ from an adult organism. The most remarkable experiment of this kind is the removal of the crystalline lens from the eye of the salamander. Now the lens of the eye develops from the primitive integument (ectoderm) of the head, but the iris of the eye develops mainly from a part of the primitive brain. After the operation a new lens is formed *from the iris* and not from the cornea. Therefore the highly specialised iris contains also determinants of other kinds. Does it contain those for itself and lens only, or others ? If it contains many kinds, then we conclude that even the definite adult structures contain determinants of many other kinds than their own, that is, reserve determinants are handed down in all cells capable of restitutive processes, practically all the cells of the body. Or does it contain only its own and those of the lens? Then this highly artificial operation was anticipated, an absurd hypothesis which need not be considered.

This particular mechanistic process (and no other one is nearly so plausible) crumbles away before attempts at verification, and it survives only by the addition of subsidiary hypothesis after hypothesis. In itself this demonstrates that it is an explanation incompetent to describe the facts.

What, then, is the "organisation"? It is something elemental, and we may just as well ask what is gravity, or chemical energy, or electric energy. It cannot be said to be any of these things or any combination of them. "At present," says a skilful and distinguished experimenter, T. H. Morgan, "we cannot see how any known principle of chemistry or of physics can explain the development of a definite form by the organism or a piece of the organism." "Probably we shall never be able," concludes Morgan, who is anything but a vitalist. But does not this mean just that in biology we observe the working of factors which are not physico-chemical ones?

We have seen that the physiologist studies something very different from that which the embryologist or naturalist studies. The former investigates a *part* of the animal, arbitrarily detached from the whole because the complexity of the functions of the simplest organism is such that all of them cannot be examined at once. He adopts the methods of physical chemistry in his investigation and whatever results he obtains are necessarily of the same order. Inevitably, from the mere nature of his method, he can see, in the organism, only physico-chemical phenomena. The embryologist, on the other hand, studies the organism as a whole and seeks to determine how definite forms are produced, and how a change in the external conditions affects the assumption of these forms. We have seen with what little success the attempts to relate embryological processes with physico-chemical ones alone have met. In all studies of organic form mechanism has failed. It is useless to attempt to press the analogies of crystalline form, and the forms assumed in nature by dynamical geological agencies. If the reader examines these analogies critically he will see that they are superficial only.

We seem, however, to see in those actions of the organism which are called " tropistic " or " tactic," reactions of a purely physico-chemical nature, and starting with these as a basis a plausible theory of organic movements on a strictly mechanistic basis might be built up.[1] A " tropism " is the movement of a fixed organism with respect to a definitely directed external stimulus. This movement may be that produced by growth of its parts, or by the differential contraction or expansion of its parts. A " taxis " we may call the motion of a freely-moving organism in response to the same directed stimuli. The movements whereby a green plant turns towards the light are called heliotropic, and those of its roots in the perpendicular direction are called geotropic. The motion of the freely-moving larva of a barnacle, for instance, in swimming towards a source of light are called " phototactic."

In all these cases we have to think of the stimulus as a " field of energy " in the sense in which physicists speak of electric, or magnetic, or electromagnetic, or

[1] Many of Jacques Loeb's remarkable investigations point in this direction.

thermal, or gravity fields. In all these cases the factors affecting the movements of the organism are directed ones.

An electric field, for instance, (1), is produced by placing the electrodes of a galvanic cell at opposite extremities of a water-trough : we imagine the electrons moving from one side of the trough to the other in parallel lines, and in a certain direction. A light field (2) would be produced by the radiation of light travelling in straight lines through the water.

The movements of the organism displaying a

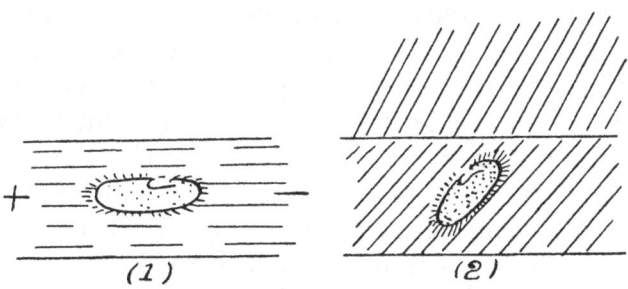

(1)   (2)

FIG. 18.

tropism or a taxis are not caused by the stimuli of the field, but are only directed by it. In the absence of these stimuli it would swim at random. In a field, however, it will orientate itself in some direction with reference to the lines of force. A " positively photo-tactic " animal swims towards the focus from which the light radiation emanates, and a " negatively phototactic " one swims in the other direction. On the theory of tropistic and tactic movements this orientation is produced by the differential stimulation of the opposite sides of the organism. Let us take as a concrete example the case of a caterpillar which creeps up the stem of a plant to feed on the tender shoots near the apex. The animal possesses an

K

elongated body, with muscles beneath the integument, and sensory nerve-endings in the latter. Its muscles are in a state of "tone," that is, they are normally always slightly tense. The incident rays of light affect the dermal sense-organs, stimulating ganglionic centres and setting up efferent impulses which descend to the muscles. Let us suppose the animal is moving so that the longitudinal axis of its body is at an angle, say of $45°$, to the direction of the incident light : one side of the body is therefore stimulated and the other is not. The stimulation of the lighted side sets up efferent nerve impulses which descend to the muscles of this side and increase their tone (or else the lack of stimulation of the other side produces impulses which inhibit the muscular tone, or impulses which would otherwise preserve the tone cease in the absence of light stimulation). In any case the muscles of the lighted side contract, and the body of the caterpillar moves so that it sets itself parallel to the direction of the radiation. Both sides of the body are then equally stimulated and the animal moves towards the light.

The animal feeds and it then creeps back down the plant. Why does it do this ? Because, says Loeb, the act of feeding has reserved the " sign " of the taxis. Before, when it was hungry, it was positively photo-tactic, but the act of feeding (all at once, it would appear, before digestion and assimilation of the food itself) has produced chemical substances in the muscles which cause the latter to relax in response to an impulse which previously produced contraction.

The nervous link is not, of course, a necessary one. The stimulation by the energy of the field may affect the muscle substance directly, or it may, as in the case of a protozoan animal, affect the general body proto-plasm in the same way. In the majority of cases, however, the orientation would be affected through the chain of sense-organ, afferent nerve, nerve centre, efferent nerve, and effector organ. This is the chain of events which on this hypothesis causes a moth to fly into a flame, or a sea-bird to dash itself against the lantern of a lighthouse.

A taxis is, then, an inevitable response by movement in a definite direction, to a directed stimulus. Including also tropisms it may be admitted that the movement is a purposeful, or at least, a useful one in some cases, as for instance the heliotropism and geotropism of the green plant. If we admit that Loeb's description of the feeding of the caterpillar, as a tactic act, is true, we may also call this a useful act. But in the majority of cases tropisms and tactes are acts which appear to be of no use to the organism. The invasion of a part of the body which is irritated by a poison (as in inflammation) by leucocytes, is useful to the body itself, but we must regard the leucocytes as organisms, and their tactic motion leads to their destruction, and so also with other analogous acts. Just because of this we find difficulty in accounting for their origin in terms of natural selection.

This does not matter so much, since it can hardly be maintained now that the tropistic or tactic act has any reality except in a very few cases—the motions of plants, galvano-taxis, the chemico-taxic movements of bacteria and leucocytes, and some other analogous cases, perhaps, are these exceptions. It can hardly be doubted that the extension of the concept to cover the motions of many invertebrates, and even some vertebrate actions, by Loeb and his school is a straining after generality which has not been justified. The hypothesis, as Loeb has stated it, is evidently almost certainly a logical one and was obviously elaborated as a protest against the anthropomorphism which saw in the flying of a moth into a flame the expression of an emotion ; or in the movements of a caterpillar on a green shrub the expression of hunger and satiety and of the inherited experience of the animal ; or in the avoidance by a *Paramœcium* of a drop of acid the emotion of dislike of the feeling of pain. Well, let it be granted that this is so, and that the protest was a useful one, for it is obviously impossible that these notions as to the causes of the movements can be verified : does it improve matters to take refuge in an hypothesis which is just as purely physico-chemical dogmatism as the other is anthropomorphism ? But the former hypothesis is at all events one which is susceptible of experimental verification and in this lies its usefulness, inasmuch as it has stimulated investigation. It is evident, however, that this verification has not yet been made. The differential afferent impulses set up by the energy-field ; the increases or inhibition of muscular tone ; the presence of photosensitive substances in the tissues of tactically acting lower animals ; the change of velocity of chemical reaction, in these cases, which ought to follow stimulation—all these things *could* be verified if they possess reality. Yet it is only indirect proofs, capable perhaps of other interpretations, and not direct experimental ones, which have so far been adduced in favour of a general theory of tropisms.

Moreover, the close analysis of the actions of some of the lower organisms by Jennings has shown that the tactic hypothesis is probably false in the majority of cases. This observer studied the acting of the organisms themselves and not the beginning and end of the series, and he shows that the behaviour of the organisms is far more obviously described by saying that it adopts a method of "trial and error." Let us suppose a number of infusoria (*Paramœcium*) in a film of water, at one part of which is a drop of acetic acid slowly diffusing out into the surrounding medium. There is a zone of changing concentrations round the drop: if we draw imaginary contours through the points where the concentration is approximately the same (the concentric rings in the diagram), and then draw straight lines normal to these rings (the radial lines) we can construct a "field" analogous to an electric or magnetic field. The animal on approaching the field ought to orientate itself and take the direction of the "lines of force." It does not, however, behave in this way, but only enters the field at random. Having entered, it remains within a part where the concentration is within certain limits. If it approaches the margin of this limited field it stops, swims backwards, revolves round its own axis, and then turns to the aboral side; and it repeats this series of movements whenever it approaches (by random) a region where the concentration is too high, or one where it is too low. In this, and other organisms we see then what Jennings has called a typical "avoiding reaction," the precise nature of which depends on the "motor-system" of the animal. Its general movements are random ones, but having found a region of "optimum conditions" (conditions which are most suitable in its particular physiological state), it remains there.

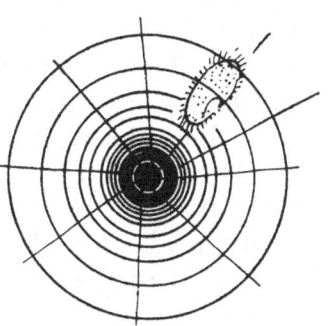

FIG. 19.

Suppose (what indeed repeatedly happens) that an

extensive " bed " of young mussels forms on a part of the sea bottom. In a short time the bed becomes populated by a shoal of small plaice feeding greedily on the little shellfish. In their peregrinations the fishes must repeatedly pass out beyond the borders of this feeding-ground. Usually, however, they will return, for failing to find the food they like they swim about in variable directions and so re-enter the shellfish bed.

Suppose (this was really a fine experiment made by Yerkes) a crab is confined in a box from which two paths lead out but only one of which leads to the water. The animal runs about at random, finds the wrong path, retraces it, tries again and again, and then finds the right path and gets back to the water. If the experiment is repeated the animal finds the right path again with rather less trouble, and after many trials it ends by finding it at once on every repetition of the experiment.

All this discussion of concrete cases leads up to our consideration of the modes of acting in the higher organisms. On the strictly mechanistic manner of thinking the actions of the organism in general are based on reactions of the tactic kind—inevitable reactions the nature of which is determined, and which follow a stimulus with a certainty often fatal to the organism displaying them. Accepting these tactic reactions as, in general, truly descriptive of the behaviour of the organism, we can build up a theory of instincts. In their simplest form instincts are reflexes—tactic movements. In their more complex forms they are concatenated reflexes, or tactes. A complicated instinctive action is one consisting of many individual actions, each of which is the stimulus for the next one ; or, of course, it may also be complex in the sense that several simple reactions proceed simultaneously, upon simultaneous stimulation of different receptors. Now the extension of all this to movements of a " higher " grade is obvious.

Let us note in the first place, that the stimuli so far considered in all the examples quoted are simple elemental ones. There are, of course, relatively few such stimuli : gravity, conducted heat (the kinetic energy of material bodies), radiated heat (the energy of the ether), electric energy, chemical energy, and mechanical contact or pressure (including atmospheric vibrations). In all these cases we have a definite, measurable, physical quantity, with which we must relate a definite response in the form of a definite measurable, physico-chemical reaction. There should be a functionality between the stimulus and response, a definite, quantitative energy-transformation. To take a concrete example, a certain quantity of light energy falling upon the receptor organs of Loeb's caterpillar ought to transform into another quantity of " nervous energy," and this travelling in an analogous way to a " wave of explosion " ought to transform into an energy quantity of some kind, which initiates another " wave of explosion " in the muscle substance. All these transformations must be quantitative ones, and the energy of the individual light must be traced from the receptor organ to the points in the muscle where it disturbs a condition of false equilibrium in the sub-stance of the latter. Nothing less than this is required to demonstrate the purely physical nature of a reaction, on the part of the organism, to an external stimulus. It may safely be said that physiological investigation has not yielded anything even approximating to such an experimental demonstration.

What are the stimuli to the actions of a higher

organism ? It is true that their elements are energies such as we have indicated, but these energies are integrated to form *individualised stimuli* (Driesch). The stimulus in an experimentally studied taxis is, perhaps, a field of parallel pencils of light rays of definite wave length ; but in the action of a man, or a dog say, the stimulus is an immensely complicated disturbance of the ether, producing an *image* upon the retina of the animal. A sound stimulus employed in an investigation may be the relatively simple atmospheric disturbance produced by the sustained note of a syren or violin-string ; but the stimulus in listening to an orchestra may consist of dozens of notes, with all their harmonies, sounding simultaneously at the rate perhaps of some hundred or two in the minute. All these are *integrated* by the trained listener, and one or two false ones among the multitude may entirely spoil the effect of the execution. Surely there is here something more than a mere difference in degree.

More important still is the strict functionality between stimulus and action that the theory of tactic responses imposes on itself. Putting this very precisely (but no more precisely than the theory demands), we say that $\Sigma A = f(x, y, z)$, that is, the series of actions ΣA (the dependent variable) is a mathematical function of the independent variables x, y, z. Now is there anything like this functionality between the acting of the higher animal and the stimulus ? Evidently there is not. We recognise someone whom we know very well by any one of a hundred different characters, mannerisms of walk, speech, dress, etc. He or she is the same person, whether seen close at hand, or afar off, or sideways, or in any one of almost infinitely different attitudes, and we respond to each of these very different physical stimuli by the same reaction of recognition : pleasure, dislike, avoidance, greeting, or whatever it may be. To a sportsman shooting wild game the stimulus may be some almost imperceptible tint or shading in cover of some kind, differing so little from its environment as hardly at all to be seen, yet, to his experience, upon this almost infinitesimal variation of stimulus depends his action with all its consequences. In Driesch's example two polyglot friends met and one says to the other, " My brother is seriously ill," or " Mon frère est sévèrement malade," or " mein Bruder ist ernstlich erkrankt " Here the physical stimulus is fundamentally different in each case, but the reaction—the expressions of sympathy and concern, the discussions of mutual arrangements, etc., are absolutely the same. Or let the one friend say to the other, " My mother is seriously ill," and in spite of the very insignificant difference between the consonantal sound *br* in this sentence and the corresponding sound *m* in the other English sentence, the reaction, that is, the subsequent conversation, and the arrangements between the two friends may be entirely different.

Putting this argument in abstract form we may say, generally, that two stimuli, which are, in the physical sense, entirely different from each other, may produce absolutely the same series of reactions ; and conversely two stimuli differing from each other in quite an insignificant degree may produce entirely different reactions. It is also easy to see, by analysis of the antecedents to the actions of the intelligent animal, that these stimuli are, in the majority of cases, not elemental physical agencies, but individualised and integrated groupings of these agencies ; and that the animal reacts, not to their mathematical sum, as it should do on a purely mechanistic hypothesis of action, but to the typical wholes which are expressed in these groupings.[1]

It is no answer to this argument to say that it is not the actual atmospheric vibrations (in the case of the conversation), nor the optical image (in the case of the recognition of a friend), which are the true stimuli, but rather the mental conditions, or states of consciousness, aroused by these physical agencies. If we are to adopt a strictly mechanistic method of explaining actions, such a method as that indicated by Loeb's hypothesis of the purely tactic behaviour of his caterpillars, then these atmospheric vibrations and optical images are most undoubtedly the true stimuli, and the reactions must be functions of them in the mathematical sense. But since this strict functionality does not exist in any behaviour-reaction closely analysed, we must grant at once that it is, indeed, *not* the physical series of events that determines the actual response, but truly the conscious state immediately succeeding to these physical sense-impressions. Now let us see to what conclusions this admission leads us.

Between the external stimulus (the atmospheric undulations impinging on the auditory membranes, or the light radiations impinging on the retinæ) and the behaviour-reaction something intervenes. This is the individual history of the organism, the " associative memory " of Jacques Loeb, the " physiological state " of Jennings, the " historical basis of reacting " (historische Reaktionsbasis) of Driesch, or the " duration " of Bergson. The last concept is the most subtle

[1] Thus to the ordinary woman the sight of a cow in the middle of a country road produces a certain definite feeling of apprehension, which is always the same although the optical image of the animal differs remarkably in different adventures.

and adequate one and we shall adopt it. The physical stimulus, then, leads to a state of consciousness, a perception, and this is succeeded by the action. What is the perception ? There may be no perception in a reflex action ; there is none in a taxis.[1] These kinds of reaction follow inevitably from the nature of the stimulus—depend upon the latter, in fact ; but we cannot fail to observe that the intelligent behaviour of the higher animal involves choice between alternative kinds of action. The perception, then, *is* this choice, or it is intimately associated with it. But it is something more than the choice of one among many kinds of response. The whole past experience of the animal enters into the perception, or at least all that part of the past experience which illuminates, in any way, the present situation. What the intelligent animal does in response to a stimulus depends not only on the stimulus but on all the stimuli that it has received in its past, and on all the effects of all those stimuli. Into the perception that intervenes between the external stimulus, then, and the action by which the animal responds what we usually call its memory enters. Its *duration* is really the something which is changed by the stimulus, and which then leads to the behaviour-reaction.

Duration, then, is memory, but it is more than memory as we usually think of this quality. The past endures in us in the form of " motor habits," and when we recall it we may act over again those motor events. Careful introspection will readily convince the reader that in recalling a conversation he is really *speaking inaudibly*, setting in motion the nerves and muscles

---

[1] We do not find this explicitly stated in this way in mechanistic biological writings. None the less it is implied, and is the legitimate conclusion from the arguments used.

of his vocal mechanisms. Actions that have been learned endure ; in some way cerebral and spinal tracts and connections become established and persist : undoubtedly when a cerebral lesion destroys or impairs memory it is these physical nerve tracts and cells that become affected. But in addition to this we have pure memory (Bergson's " souvenir pur "). What, for instance, is the visual image of some thing seen in the past, which most people can form, but pure recollection ? [1]

All the past experience of the organism—all its perceptions, and all the actions it has performed— endures, either as motor habits or mechanisms, or as pure memories. All this need not be present in its consciousness ; the motor habits would not, of course, and only so much of the past would be recalled as would be relevant to the choice which the organism was about to make of the many kinds of responses possible to its motor organisations. Out of this past it would select all that was connected in any way with the actions which were possible to it in the present. It would recall all actions previously performed which resembled the one provisionally decided upon ; but recalling also the other circumstances associated with those past actions, it would discover something which would lead it to modify that provisional action. Now in describing the whole behaviour of the acting organism in this way are we doing any more than simply expressing in more precise terms the " commonsense " notions of the ordinary person ? The latter would sum up all this discussion by saying that what he would do in any set of circumstance depended not only on the circumstances themselves but upon his experience.

---

[1] A visual image may, of course, be something that has never been actually seen. But then its elements have had actual perceptual existence in the past.

Physiology shows us as clearly as possible that in the stimulation of a receptor organ, the propagation of a nervous impulse along an afferent nerve, the transmission of this impulse through the cord or brain, or both—in the propagation again of the impulse through an efferent nerve and the transformation of this impulse into a releasing agency, setting free the energy potential in the muscle substance—that in all this there can be nothing more than physico-chemical energy-transformations. All this is clear and certain. But why should the same afferent stimuli, entering the central nervous system at different times by the same avenues, and in the same manner, traverse different tracts, and issue along different efferent nerves, producing different results ? Or why should different stimuli entering the central nervous system take the same intra-cerebral paths and then affect the same efferent nerves and effector organs ? It is because these stimuli lead to perceptions which fuse with, and become part of the duration of, the organism. And the response then becomes a response not to the physical stimulus, but to the duration modified in this way.

Can we conceive of any physical mechanism in which the duration of the organism accumulates ? Can we think of any way in which memories are stored in the central nervous system ? When we say " stored," it is our ingrained habit of thinking in terms of space and number that makes us regard memories as laid by somewhere, in the way we file papers in a cabinet, or store specimens in a museum. Supposing perceptions are stored in this way, we think of them as stored or recorded in the same way as a conversation is recorded and stored in a phonograph. The phonograph can reproduce the conversation just as it was received, but what we make use of when we utilise our experience is obviously the elements of that experience, selected and re-integrated as we require them. There must, then, be something like an analysis of our perceptions, a dissociation of these into simple constituents, and a means of restoring and recording these constituents in such a way that they can be re-combined in any order, and again made to enter into our consciousness.

It is quite possible to imagine such a mechanism. Let us suppose that an efferent impulse enters the cerebral cortex *via* any one axon: there is a perfect labyrinth of paths along which the impulse may travel. Everywhere in the central nervous system we come upon interruptions of nervous paths formed by inter-digitating arborescent formations. The twigs of these arborescences do not, apparently, come into actual contact with each other and the impulse leaps across the gap between them. This gap is, of course, ex-ceedingly narrow, and one can almost speak of it as a membrane, since it must be occupied by some organised substance. It has been called the synaptic membrane. Let us suppose that a stimulus of a certain nature passes through the synapse, modifying it physico-chemically as it passes. Thereafter a stimulus of similar nature will tend to pass across this particular synapse, the resistance of the latter having been decreased. It will thus tend to travel by a definite tract through the central nervous system. Now the latter we may regard in a kind of way as a very com-plicated switchboard, the function of which is to place any one stimulus (or series of stimuli) out of many in connection with any one motor[1] mechanism (or series

[1] Or more generally *effector* mechanism. This enables us to include reactions, such as secretory ones, which are not motor.

of mechanisms) out of many. A motor habit, or path, is then established and will persist.

Such a conception is clear and reasonable in principle, and all work on nervous physiology tends to show that it is a good working hypothesis. We cannot read modern books without feeling that immense advances will be made by its aid. But the complexity of the brain of the higher vertebrate is so incredibly great, and the difficulties of imagining the nature of the necessary physico-chemical reactions in the synapses, and elsewhere, are so immense that experimental verification may be impossible. And all that we have said applies to a single elemental stimulus, yet in any common action the stimulus is a synthesis of almost innumerable simple ones, while the response is also a synthesis. The optical image of almost any object contains a very great number of tints and colours differing almost imperceptibly : there must at least be as many simple stimuli as there are rod or cone elements in the part of the retina covered by the image. The motor responses consist of a multitude of delicately adjusted and co-ordinated muscular contractions and relaxations. If we are to accept a mechanistic hypothesis of action, of this kind, and which includes only such processes as are suggested above, it is not enough that a logical description, consistent in itself, and consistent with physico-chemical knowledge, should be formulated. The mere statement of such an hypothesis does not carry us far. If it is, in essence, mechanistic, it must be capable of experimental verification in detail.

Even if it were verified experimentally it would still leave untouched the problem of consciousness. All that we have considered are series of physico-chemical energy-transformations. How, then, does consciousness arise ? We cannot even imagine its association in a functional sense with the train of events forming an afferent impulse. In some form or other mechanism must assume a dualism—a parallelism of physical and psychical processes. Physical events in the central nervous system are associated with psychical ones—when the former occur so do the latter —yet the former are not "causes" in any physical sense of the latter. Consciousness follows cerebral energy-transformations as a parallel " epiphenomenon." At once we leave the province of mechanism, and how can we remain content with such a limitation of our descriptions ? And if we conclude, as we seem obliged to do, that consciousness is an affective agency in modifying our responses to external stimuli, does not this in itself show that our concept of behaviour as a purely physico-chemical process is insufficient in its exclusiveness ?

We return to a consideration of the main results of experimental embryology in a later chapter, but let us notice here what is the direction in which these results, and those of the analysis of instinctive and intelligent action, carry us. It is towards the conclusion that physico-chemical processes in the organism are only the *means* whereby the latter develops, and grows, and functions, and acts. In the analysis of these processes we see nothing but the reactions studied in physical chemistry ; but whenever we consider the organism as a whole we seem to see a co-ordination, or a control or a direction of these physico-chemical processes. Nägeli has said that in the development of the embryo every cell acts as it if *knew* what every other cell were doing. There is a kind of autonomy in the developing embryo, or regenerating organism, such that the normal, typical form and structure comes into existence even when unforeseen interference with the usual course of development has been attempted : in this case the physico-chemical reactions which proceed in the normal train of events proceed in some other way, and the new direction is imposed on the developing embryo by the organisation which we have to regard as inherent in it. This same direction and autonomy must be recognised in the behaviour of the adult organism as a whole. What is it ? We attempt to think of it as an impetus which is conferred upon the physico-chemical reactions which are the manifestations of the life of the organism. It is the *élan vital* of Bergson, or the *entelechy* of Driesch. What is included in these concepts we consider in the last chapter of this book ; and before so doing it will be necessary to consider the organism from another point of view, that of its mutability when it is regarded as one member of a series of generations.

L

CHAPTER V

WHAT is an individual organism? A Protozoan, such as an *Amœba* or a *Paramœcium*, is a single cell : it is an aggregate of physical and chemical parts, nucleus, cytoplasm, etc., and no one of these parts can be removed if the organism is to continue to live. The cell can be mutilated to some extent, but, in general, its life depends on the integrity of its essential structures, and it cannot be divided without ceasing to be what it was. It contains the minimum number of parts which are necessary for continued organic existence.

Such an organism as a *Hydra* consists of an aggregate of cells which are not all of the same kind. The outer layer, or ectoderm, is sensory and protective, and contains organs of aggression ; while the inner layer consists of cells which subserve the functions of digestion and assimilation. All these parts are, in general, necessary for the life of the *Hydra*. They can be mutilated ; the animal can be cut into two parts, and each of these parts may regenerate, by growth, the part that was removed. Yet the existence of ectoderm and endoderm, in a certain minimum of mass, is necessary for this regeneration. The higher animal, or Metazoon, is therefore an aggregate of cells, each of which is equivalent to the individual Protozoon ; but these cells are not all alike—that is, there is differentiation of tissues in the multicellular organism.

Again, the Cœlenterates provide examples of animals which are aggregates of parts, each of which is the morphological equivalent of a single *Hydra*. Such an animal as a Siphonophore, for instance, consists of zooids, and each of these units has the essential structure of a *Hydra*. But the zooids are not all alike : some of them subserve the function of locomotion, others of aggression, others of digestion and assimilation, and so on. Here, again, the whole organism may be mutilated ; parts may be removed and regeneration may occur ; but, as a Siphonophore, all of the different zooids must be present if the characteristic functioning of the animal is to continue.

The Protozoon is, therefore, an individual of the first order, the *Hydra* an individual of the second order, and the Siphonophore an individual of the third order. Some such conception of degrees of individuality will probably be regarded as satisfactory by most zoologists, yet consideration will show that it is very inadequate. Many unicellular plants and animals may consist of a number of cells, which are all alike. The Diatoms and Peridinians reproduce by the division of their cell bodies and nuclei, and the parts thus formed may remain in connection with each other. A Diatom may consist of one cell, or it may consist of a variable number of such connected together by filaments, or in other ways ; and the dissociation of such a series may occur without interfering in any way with the functioning of the parts separated. A Tapeworm consists of a " head " or scolex, containing a central nervous mass and organs of fixation ; and organically continuous with this is a series of segments or proglottides. These proglottides are formed continuously from the posterior part of the scolex, and they may remain in connection with each other, and with the central nervous system and some other organs which are concentrated in the scolex. Nevertheless, each proglottis contains a complete set of reproductive organs ; it has locomotory organs so that it can move about, and can fix itself to any surface into which it comes in contact. It can lead, for a considerable time, at least, an independent existence apart from that of the scolex and the other proglottides with which it was originally in continuity. In the majority of Polyzoa, the common Sea-Mat, for instance, the organism consists of a very large number of polypes or zooids, each of which secretes an investment of some kind round itself, but all of which may be connected together by a common flesh. In many Zoophytes there is essentially the same structure. In Corals there are very numerous zooids, each of which lives in a calcareous calyx secreted by itself. Polyzoa, Zoophytes, and Corals are individuals of the third order, and we might regard the tapeworm strobila— the scolex with its chain of proglottides—as belonging also to the same category. Nevertheless, a part of a Polyzoan or Hydrozoan colony, or a proglottis from a tapeworm, may become detached, when it will continue to live and reproduce and exhibit all the characteristic functioning of the species to which it belonged.

Such an animal as a *Hydra*, or a Planarian or Chætopod worm, or a starfish, may be cut into several pieces, and provided that each of these pieces exceeds a certain minimum of mass, it will regenerate the whole structure of the organism of which it formed a part. In the developing embryo of the Sea-urchin the eight-cell stage may be treated so that the blastomeres may come apart from each other : each of them will then begin to segment again and will reproduce the typical larval Sea-urchin. The parasitic flat-worm, known as the liver-fluke, produces larvæ which develop to form other larvæ called rediæ. Each redia normally develops into another larval form, called a cercaria, which finally develops into the adult worm. But in certain circumstances each redia may divide and reproduce a number of daughter-rediæ, and there may even be several generations of these larvæ. In many lower animals buds may be formed from almost any part of the body, and each of these buds may reproduce the entire organism. In plants the entire organism may be grown from a very restricted part or cutting. Thus the individual, whether of the first, second, or third order, may be divided without necessarily ceasing to be what it was.

Regeneration of fragments detached from the fully developed adult body so as to form complete organisms does not, in general, occur among the higher animals, nor, as a general rule, does reproduction by bud-formation occur. When such animals reproduce, an ovum develops to form a large mass of cells, which later on become differentiated to form the tissues and organs of the adult body. But a relatively small number of the undifferentiated cells persists in the ovaries of the females, or in the testes of the males, and each of these cells may again develop and reproduce the organism. There is apparently no limit to this process : any animal ovum may become divided successively so that an infinite geometrical series is produced, and in every term of this series all the potentialities of the first one are contained.

The physical concept of individuality—that which cannot be divided, or which may not be divided without ceasing to be what it was—such individuality as the chemical molecule possesses cannot be applied to the organism. Any definition that involves the idea of materiality fails. Obviously the notion of the individual most commonly met with in zoological writings—that it is the product of the development of a single ovum—fails, for, logically applied, it would regard the entire progeny of the ovum, that is, all the organisms belonging to the species, as the individual. It is clear that the difficulties of the concept arise from our attempt to identify the life of the organism with the material constellation in which this life is manifested. In the course of generation after generation the ovum becomes divided and grows and is again divided, and so on without apparent limit. But if we assume that the " organisation " or " entelechy " is material and is capable of this infinite divisibility without impairment of its attributes, do we not extend to matter a property which belongs only to the concepts dealt with by mathematics ?

The discussion of individuality with regard to the organism, considered as a morphological entity, is, indeed, rather a formal one, and it is valuable only in so far as it has for its object the establishment of the most convenient terminology. Nevertheless, the notion of organic individuality is clear to us though it is a notion felt intuitively and incapable of analysis. We see in nature animals like ourselves, and we do not doubt that each of them is an entity isolated from the rest of the universe, and to which the rest of the universe is relative. We ourselves are primarily centres of action. Motion, or change of position with respect to some object apart from ourselves in nature, is only relative, and there is no standard or point in the universe which is motionless and to which we can refer the motion of a body apart from our own. But the motion of our own body is something felt or experienced intuitively, something absolute. As we move, the universe, our universe rather—that is, all that we *act upon, actually or in our contemplation* —contracts in one direction and expands in another. We feel ourselves to be apart from it although we may, to some extent, control it. We have no doubt that the higher animals have this feeling of isolation from, and relation to, an universe which is something apart from themselves ; though, of course, the attempt to demonstrate this leads to all the kinds of difficulties suggested in our attempt to discuss individuality. It is a conviction so strongly felt that we have no doubt about it. The organic individual we may then describe as an isolated, autonomic constellation of physico-chemical parts capable of indefinite growth or reproduction.[1]

What is reproduction ? It is organic growth by dissociation accompanied in the higher organisms by differentiation and reintegration. To make this statement clear, we must now consider the phenomena of reproduction in the lower and higher organisms.

We know purely physical growth. If a small crystal of some suitable substance be suspended in an indefinitely large quantity of a solution of the same chemical substance it will begin to grow, and there is no apparent limit to the mass which it may attain. Such giant crystals may be grown in the laboratory or they may be found in rock masses. Growth here is a process of accretion in which a particular form is maintained. Form in inorganic nature may be essential or accidental. Accidental forms are such

[1] The description is, of course, only a convenient one. The notion of individuality, as it is expressed in the earlier part of this paragraph, is an intuitively felt, or subjective, one. It is best called personality.

as are partially the result of a very great number of small and unco-ordinated causes : the form of an island or a mountain suffering erosion, or the shape of a river valley or delta, or the arrangement of the stones forming a moraine at the side of a glacier. Essential forms are such as are assumed as the result of the operation of one or a few co-ordinated causes, and such are the forms of crystals. They are invariable, or they vary within very small limits about an invariable mean form.

The form of a crystal depends on the structure of the molecules of the chemical substance from which it is produced. We cannot, of course, speak of the shape of a molecule, but we know that the atoms of which it is composed have certain positions in space relative to each other—positions which are conceptualised in the structural formulæ of the chemists. In the solution, or mother-liquor, these molecules move freely among each other, but in the crystal they become locked together and their motions are restricted. The shape of the crystal depends on the way in which the molecules are locked together, or on the way in which they are arranged. A cube may be built up by the arrangement of a number of very small cubes : obviously we could not make a cube from a number of very small hexagonal prisms if the latter were to be packed together in such a way as to occupy the minimum of space. An infinitely great number of cubes might also be formed by adding single layers of very small cubes to the faces of an already existing one—that is, by the accretion of elements of essentially similar form. In every cube (or crystal) of this infinite number the geometrical form would be the same, and if we were to measure any one side of any cube of this series we should find that the total surface would always be a definite function of the length of this side. The mass of a cube would also be a function of such a measurement : it would be al^3, a being a constant depending on the unit of mass and on the specific weight of the substance of which the crystal was composed. If we take a series of crystals of increasing size, this relation holds for every one of them : $M=al^3$, M being the mass, a the constant referred to above, and l, the independent variable, being any one length of a side of the crystal.

If the organism grows by accretion in the same way as does a crystal, this relation ought also to hold in all the exclusiveness with which we expect it to hold in the growth of a crystal. But it does not so grow. Its growth is something essentially different, and none of the superficial analogies so prevalent nowadays ought to obscure this difference. The organism may grow by accretion, thus layers of calcareous matter may be added to the outside of a membrane bone from the investing periosteum, or it may grow by the deposition of matter within the actual cell bodies, (the process of growth by intussusception of the plant physiologists). But the extent of growth by accretion is strictly limited in all organisms : for each there is a maximal mass determined by the nature of the animal or plant, and this mass is that of the uni-cellular organism itself, or that of the cells of which the multi-cellular organism is composed. There may also be growth by accretion in the case of the formation of skeletal structures, which are laid down by the agency of the cells of the organism ; but if we confine our attention to the growth of the actual living sub-stance we shall see that accretion ceases when the mass characteristic of the cells has been attained, when growth by dissociation begins. The cell then divides, and each of the parts into which it has divided grows to the limiting size, and division again occurs. This is what happens in the case of the growth of the Sea-urchin egg to form the larva, or blastula. The ovum segments into two blastomeres, each of which then grows to a certain extent, and again segments into two blastomeres. After the completion of ten divisions there are about 1000 cells which are arranged so as to form a hollow ball—the blastula.

Differentiation is now set up. In the blastula stage all the cells are alike, actually and potentially. But soon one part of the hollow ball of cells becomes pushed inwards, and the cells of this inturned layer become different from those of the external layer, while cells of a third kind appear in the space between the external and internal layers.

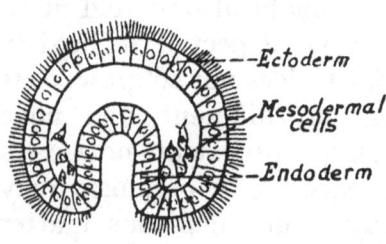

FIG. 20.—The Sea-urchin Gastrula larva in section.

This is the process of differentiation leading to the development of the various tissues— protective, sensory, digestive, skeletal, etc. The cells still continue to divide and grow to their maximal size, but when the process of differentiation begins, the cells which are formed are not quite the same as those from which they originated. Finally, however, when the rudiments of all the tissues of the adult body have been laid down, the cells begin to produce daughter-cells of only one kind. Growth of the embryo consists, therefore, of the dissociation or division of the substance of the ovum and blastomeres, followed by a gradually increasing differentiation of the cells so produced.

Reintegration proceeds all the time. Blastula and gastrula larvæ are really organisms capable of leading an independent existence—that is, they are autonomous entities or individuals. The activities of the parts of which they are composed—ectodermal locomotory cells, ectodermal sensory cells, endodermal assimilatory cells, and so on, must be co-ordinated. The cells are in organic material continuity with each other, and events which occur in any one of them affect all the rest. Impressions made upon the sensory cells are transmitted to the locomotory cells, and food-material assimilated by the assimilatory cells is distributed to all the others. At all stages the growing embryo is an organic unity. The more fully it is developed, the greater the morphological complexity of the organism, and the more numerous its activities, the greater is the differentiation ; but the greater also is the co-ordination of the organs and tissues. In the higher animals this co-ordination and integration of activities is effected (mainly) by the central and peripheral nervous systems, but specially differentiated nervous cells are not necessary for this purpose. Differentiation during growth is therefore necessarily accompanied by reintegration of the parts dissociated and differentiated.[1]

In the process of organic growth the relation between mass and form no longer holds in all the

---

[1] Societies and civilisations, the associations of bees and ants, or the Modern State, obviously exhibit this differentiation. It is morphological and functional in the case of the Arthropods, since individuals performing different duties are modified in form. It is functional only in the case of human societies. Integration of the activities of the individuals in both kinds of societies is effected by inter-communication : articulate symbols in the case of the lower animals, language in the case of man. If the concept of " orders of individuality " were anything more than a convenient, though artificial, analysis of naturally integral entities, we might speak of the ideal state or the insect society as a " fourth order of individuality."

exactness with which it applies to the growth of the crystal. We might spend a lifetime growing tablets of cane-sugar, but in all cases we should find that the mass of any crystal was proportional to the cube of a length of a diameter : there would be a strict relation between mass and geometrical form. But this strict relation does not hold in the case of a series of organisms belonging to the same species but differing in size. If we measure, for instance, the lengths of a great number of fishes of the same species, we should find that we must describe the law of growth, not by the simple equation $M = al^3$, but by an empirically evaluated expression of the form $M = a + bl + cl^2 + dl^3 +$ . . . and that the constants in this equation would vary with the species studied and with the conditions in which it is living—that is, the organism changes in form as it increases in size. This is inconceivable in the case of purely physical growth by the accretion of molecules, and we find again that the characters of the organism depend not only on what it is but also upon what it has been—that is, upon its duration. Growth, then, in plants and animals implies variability in form, in general cumulative variability, leading to an indefinite departure from the typical form.

The organism, therefore, does not grow simply by the accretion of material, but, having attained a certain limit of size, it divides or reproduces. In the lowest plants and animals this process of division is simple : either the organism (unicellular or multicellular) divides itself into two approximately equal parts or it divides into a number of such parts. The first process is represented by the reproduction of a bacterium or an *Amœba*, or by the division of a Planarian worm ; the second is represented by the division (in many Protozoa, for instance) of the whole

organism into a number of spores. Fundamentally
the two processes are alike : the simple, binary division
of the Bacterium is followed at once by growth by
accretion, while in brood-formation (the cases of
multiple division) the parent cell divides, and then
each of the daughter-cells divide, and so on for
several generations. After the completion of these
divisions the brood-cells grow by accretion to their
normal size. It is meaningless, in the light of our
previous discussion, to say that the individuality of the
mother-cell " is merged in that of the daughter-cells."
But we may believe that a *Paramœcium* possesses
some degree of consciousness. Does it possess personality—that is, the feeling of isolation from the rest of
the universe, and the feeling of oneness with its own
past-memory or conscious duration ? If so, its personality, when it divides, becomes one with that of its
daughter-cells. Or is its personality and conscious
past that also of its sister-cells, and also that of the no
longer existent mother-cell, and the cell of which this
in its turn was a part ? We must remember that such
an organism as a *Paramœcium* shows in its behaviour
most of the signs of intelligence ; that the parts into
which it divides when it reproduces are equally developed ; and that the process of division may not
interrupt the conscious duration of either part. Is
there a common personality, or oneness of consciousness, of all the organisms of this kind which are
descended from the same individual ?

Reproduction by division, simple or multiple,
does not proceed indefinitely in the case of the unicellular organisms. Sooner or later there is a limit,
and the cell is then no longer able to continue dividing. Conjugation then occurs in one of many modes.
Essentially two organisms come into contact and

their nuclei fuse, or rather some of the material of one nucleus is transferred to the other. The cells then separate and reproduction by division begins again.

This is not necessarily sexual reproduction : it is the conjugation of essentially similar morphological entities. If two conjugating *Paramœcia* possessed distinct personalities we might imagine a merging or addition of two conscious durations or memories. Sexuality, however, includes less than this. In this mode of reproduction the conjugating bodies are not organisms in the usual sense, but rather modified organisms or highly modified parts of organisms. In some lower plants the conjugating cells may be modified with respect to the cells characteristic of the organism, but they may be approximately equal in size. But in the multicellular plant and animal, in general, the conjugates are cells detached from the parental body, and differing chiefly from the cells of the latter in that they show a lack of differentiation. One of these cells, that detached from the paternal body, is the spermatozoon (in the case of the animal), or the pollen cell (in the case of the plant). It is much smaller than the sexual cell detached from the maternal body : this is the ovum in the case of the animal, or the oosphere in the case of the plant. In general the ovum is a relatively large cell, since it contains abundant cytoplasm, which may also be loaded with yolk or other reserve food material. The spermatozoon is very much smaller and consists of a nucleus with a minimal mass of cytoplasm. The ovum is, in general, immobile ; the spermatozoon is generally highly mobile.

The essential process in the sexual reproduction of the unicellular organisms is therefore the conjugation of the organisms themselves. In multicellular organisms, modified cells—the germ-cells—become detached from the bodies of the parents, and these cells conjugate. In many lower plants and animals phases of sexual and asexual reproduction alternate, thus *Paramœcium* reproduces by simple division, but at intervals conjugation occurs. In plants sporophytic and gametophytic generations alternate, the sporophyte reproducing by multiple division—that is, by the formation of spores, and the gametophyte reproducing by the formation of germ-cells. There are few organisms—possibly none—in which continued asexual reproduction by simple or multiple division, spore-formation, bud-formation, etc., can proceed without limit. In the great majority of cases investigated asexual reproduction becomes feeble after a time and then ceases, and it has been held that the stimulus of conjugation of the cells, or that of sexual reproduction, is necessary for its renewal. In such cases. the organism is said to have become " senescent," and " rejuvenescence " by some means becomes necessary. As a general rule rejuvenescence is effected by the interchange of nuclear matter between two conjugating organisms, but it may be effected by rest, or by a change of environment, or by the supply of some unusual food-material to the liquid in which the dividing organism is contained. Thus, if various materials be added to the water inhabited by a dividing *Paramœcium*, the Protozoon may continue to reproduce by simple division for at least two thousand generations. We must remember, however, that " senescence " and " rejuvenescence " are only words; what is the essential nature of the changes denoted by them we do not know.

In sexual reproduction, as it occurs in the great majority of plants and animals, the ovum, or female germ-cell, is fertilised or " activated " by the male germ-cell. But this activation by the spermatozoon is not necessary, for the ovum itself is capable of division and development to form a complete organism. This occurs in the cases of natural parthenogenesis among insects and some other animals, where the ovum proceeds, without fertilisation, to segmentation and development. In some lower plants, where the size of the male and female germ-cells is nearly equal, either of them may undergo parthenogenetic development : in such cases we cannot, of course, properly speak of sexual differentiation. In the cases of organisms normally reproducing sexually, the stimulus to development is afforded by the entrance into the ovum of the spermatozoon—that is, by the mixture of the male and female germ-plasms ; but in some animals this stimulus may be replaced by the addition to the water in which they are living of certain chemical substances. This is the process of artificial parthenogenesis first studied by Loeb in the case of the eggs of the Sea-urchin ; and its analysis suggests that the spermatozoon conveys some substance into the egg, and that this substance initiates segmentation by setting up a train of chemical reactions. What these reactions are exactly, and what is the process of " formative stimulation " by the spermatozoon, we do not know. It is quite certain, however, that much more than this process of formative stimulation is involved in the fertilisation of the egg by the spermatozoon. The mixture of the male and female germ-plasms resulting from conjugation confers upon the embryo the characters of both the parents and of their ancestries.

In an unicellular organism the " body " consists of a single cell containing a nucleus. The extra-nuclear part of the cell—the cytoplasm—is modified in various ways : thus it may possess flagella, or cilia, so that it may be actively locomotory. It is at once a receptor apparatus, susceptible to changes in the medium in which it lives, and it is also an effector apparatus, capable of transforming stimuli received into motor impulses. It may be able to accumulate available energy by making use of the energy of radiation in the synthesis of carbohydrate and proteid from the inorganic substances in solution in the water in which it lives ; and it is also able to expend this energy in controlled movements. All the characteristics of life, in fact, are exhibited by the unicellular organism, the differentiation of the cytoplasm corresponding functionally to the differentiation of the tissues of the multicellular animal or plant.

In the latter the organs, organ-systems, and tissues are composed of differentiated cells. Development consists essentially of a process of cell-formation by simple division, and at the end of this process of segmentation various rudiments (Anlagen) are established. The older embryologists sought to recognise the formation of three " germ-layers " in most groups of animals : these were the outer layer or ectoderm, the middle layer or mesoderm, and the internal layer or endoderm. The ectoderm, it was held, gave rise to the integument, the central and peripheral nervous systems, and the sensory organs. The mesoderm gave rise to the musculature and skeleton, the excretory organs, and some other tissues. The endoderm gave rise mainly to the alimentary canal and its glands. The " Gastrea-Theory " of Haeckel sought to recognise a similar larval form, or " Gastrea," in the development of most multicellular animals, and much ingenuity of argument was required for the estab-

M

lishment of this homology. The newer embryology recognises the difficulties implied in the application, in all its exclusiveness, of the Gastrea-theory to the higher phyla of multicellular animals ; so that nowadays it has been necessary to abandon the notion of the metazoan animal as being built up from these three primary germ-layers. At the conclusion of segmentation, then, the embryo consists of a mass of cells similar to each other in structure, but differing in fate and in potency. Some of these cells are destined to give rise to the integument, the nervous system, and the sense-organs ; others become the skeleton and musculature ; and others again the organs of digestion, assimilation, and excretion. A primary arrangement of these groups of cells into three layers is indeed set up in many cases of development, but it is plain that this arrangement is far from being an universal one. Modern embryology shows in the clearest possible manner that at the end of segmentation the embryo consists of a group of cells each of which has normally a different fate in subsequent development. What precisely each cell will become depends on its position with regard to the others. But each cell is capable of becoming more than it normally becomes : its potency is greater than its actual fate. If the normal course of development is interrupted, a cell, which would usually have given rise to a part of the skeleton, may give rise to a part of the alimentary canal. The cells of the developing embryo are autonomous.

In the normal course of development most of the cells existing at the end of segmentation give rise to the " body " of the organism, undergoing differentiation as they so develop. But a few embryonic cells persist without structural modification throughout the development of the animal. They divide and grow and become greater in number, but remain unchanged in other respects. These cells become the essential reproductive organs, or gonads, of the adult animal— that is, the ovaries of the female and the testes of the male. In the females of the higher animals (the mammals, and perhaps some of the Arthropods) these cells only divide and grow during the early stages of development, and long before the beginning of adult life the number of ova in the gonads has become fixed. In all males, and in the females of most animals, however, the reproductive cells appear to be capable of unlimited multiplication.

The essential cells of the gonads, the ovarian mother-cells or the sperm mother-cells, constitute the germ-plasm. In modern, speculative, biological literature the term germ-plasm is, however, restricted to the chromatic material in the nuclei of the reproductive cells, the cytoplasm being regarded as non-essential for the transmission of the hereditary qualities of the organism. It seems clear, however, that this distinction between the cytoplasm and the chromatic matter of the nucleus is not always a valid one, so that it is best to speak of the whole cell as constituting the germ-plasm. The embryonic cells, therefore, have different fates : some of them become transformed during development into the body or *soma*, and others remain unmodified throughout life as the *germ*. The soma enters into intimate relationships with the environment ; it is affected by the vicissitudes of the latter ; and it may actively respond to them. The germ-cells may possibly migrate through the body, perhaps, it has been suggested, developing fatally and irresponsibly into the mysterious, malignant tumours of adult life. Normally, however, they remain segregated in the reproductive glands, secluded from the outer environment. Their activities are inherent in themselves, are rhythmic, and become functional only on the assumption by the soma of the phase of sexual maturity. From the point of the species the soma is only the envelope of the germ-cells. It is affected by every change of the environment, and being usually cumulatively affected by the latter it becomes at length an unfit envelope. Somatic death then follows as a natural consummation, but the germ-cells are, in a sense, immortal in that they remain capable of indefinite growth by division.

In the sexual reproduction of the higher organism a part of the germ-plasm becomes detached, undergoes growth, and develops into an organism exhibiting the parental organisation. But in the development of the offspring, part of the germ-plasm received from the parent persists unchanged, is transmitted to another generation, and so on without apparent limit. *Something is transmitted from parent to offspring.* This something we regard as a cell exhibiting a definite chemical and physical structure ; but while the germ-cell differs in certain respects from an ordinary somatic cell, these structural and chemical differences are insignificant when they are compared with the differences in the potentialities of the cells. The somatic cells are, in general, capable of reproducing only the general character of the tissues of which they form part. Some of them, the cells of the grey matter of the central nervous system, for instance, appear to be incapable of division and growth. But again the facts of regeneration appear to point to the possession by the somatic cells of more than this restricted power of reproducing the tissues of which they form part : to this extent the regeneration experiments tend to remove the essential distinction between the somatic and germinal cells. Neglecting these results in the meantime, we see that the germ-cells contain within themselves the potentiality of reproducing the entire organism in all its specificity. That which is transmitted from the parent to the offspring is the parental organisation in all its specificity ; and to say that this organisation is a material thing is, of course, to state a hypothesis, not a fact of observation.

This transmission of a specific form and mode of behaviour from generation to generation is what a hypothesis of heredity attempts to explain—that is, to describe in the simplest possible terms, making use of the concepts of physical science. " Twelve years ago," says Jacques Loeb, " the field of heredity was the stamping ground for the rhetorician and metaphysician; it is to-day perhaps the most exact and rationalistic part of biology, where facts cannot only be predicted qualitatively, but also quantitatively." Let the reader examine for himself the meagre array of facts on which this apotheosis of mechanistic biology is based.

Two modern hypotheses of heredity demand attention—Weismann's hypothesis of the continuity of the germ-plasm, and Semon's " Mnemic " hypothesis. In the latter it is assumed that the basis of heredity is the unconscious memory of the organism : modes of functioning are " remembered " by the germ-plasm and are transmitted. This notion presents many points of similarity to that which we consider later on, so that it need only be mentioned here. Weismann's hypothesis—like Darwin's hypothesis of Pangenesis— is a corpuscular one, and has obviously been suggested by the modern development of the concepts of molecules and atoms in the physical sciences. It supposes that that which is handed down is a material substance of a definite chemical and physical structure. This is not the germ-cell, nor even the nucleus of the latter, but a certain material contained in the nucleus. The latter contains protein substances containing a greater proportion of phosphoric acid than does the cytoplasm of the cells in general; these proteins are known as nucleo-proteins, though our knowledge of their chemical structure is, so far, not very exact. It is not, however, these that are the germ-plasm, but a substance in the nucleus *which becomes visible when the cell is killed in certain ways, and which becomes stained by certain basic dyes.* It is distinguished by this character alone and on that account is loosely called " chromatin." This substance Weismann identifies as " the material basis of inheritance."

When a cell divides, a very complex train of events usually occurs. This process of " Mitosis " exhibits many variations of detail, and without actual demonstration it is rather difficult to explain clearly. But its essential feature is evidently the exact halving of all the structures in the cell which is about to divide. In the ordinary cell which is not going to divide immediately, the chromatin is diffused throughout the nucleus as very numerous fine granules, recognised only by their staining reactions. They may be concentrated at some part of the nucleus, so that a division through a plane of geometrical symmetry of the cell would not, in general, exactly halve the chromatin. Prior to division, therefore, this substance becomes aggregated as granules lying along a convoluted filament of a substance called " linin," which is characterised principally by the fact that *it does not stain with the dyes that stain the chromatin.* The filament breaks up into short rods, called Chromosomes, and these rods become arranged in the equator of the nucleus. The rods then split longitudinally, and one-half of each moves towards one pole of the nucleus, the other half moving towards the other pole. Various other modifications of the cell and nucleus occur concomitantly with these changes, but the essential thing that happens seems to be the halving of all the structures of the cell, and this is the simplest explanation of the phenomena of mitotic cell division. Two daughter-cells are then formed by the division of the mother-cell, and each of these daughter-cells receives one-half of each of the chromatin granules that were contained in the mother-cell.

The chromosomes, or " Idants," are seen to consist of discrete granules, and these are (generally) the bodies known as the " Ids." The id cannot be resolved by the microscope into any smaller structures : it lies on the limits of aided vision ; but the hypothesis assumes that it is composed of parts called " Determinants," and the determinants are further supposed to consist of " Biophors." The biophors are the ultimate organic units or elements, and they are of the same order of magnitude as chemical molecules. We must suppose them to be more complex than a protein molecule, and the latter contains many hundreds (at least) of chemical atoms. Now it is possible to calculate the number of atoms contained in a particle of the same size as the id : such a calculation may be made by different methods, all of them yielding concordant results. This calculated number of atoms may be less than that which we must suppose to be present in the biophors, of which the hypothetical id is composed ! [1]

---

[1] " But," says Weismann, referring to an objection of this nature, " it should rather be asked whether the size of the atoms and molecules is a fact, and not rather the very questionable result of an uncertain method of investigation."

The id is supposed to contain all the potentialities of the completely developed organism. It is composed of a definite number of determinants, each of the latter being a " factor " for some definite, material constituent of the adult body. There would be a determinant for each *kind* of cell in the retina of the eye, one for the lens, one for the cornea (or rather for each kind of tissue in the latter), one for each kind of pigment in the choroid and iris, and so on ; every particular kind of tissue in the body would be represented by a determinant. Thus packed away in a particle which lies just on the limits of microscopic vision are representatives of all those parts of the body which are chemically and physically individualised, each of these hypothetical " factors " being a very complex assemblage of chemical atoms. In development the determinants become separated from each other, so that whatever parts of the body are formed by the first two blastomeres are represented by determinants which are contained in those cells, and which are sifted out from each other and segregated. As development proceeds this process of sifting becomes finer and finer, until when the rudiments of each kind of tissue have been laid down a cell contains only one kind of determinant. This consists of biophors of a special kind, and the latter then migrate out from the chromatin into the cytoplasm of the cells in which they are contained, and proceed to build up the particular kind of tissue required.

The nucleus of the germ-cell is thus a mixture of incredible complexity, but in addition to this material mixture there must exist in it the means for the *arrangement* of the determinants in the positions relative to each other occupied by the adult organs and tissues. A mechanism of unimaginable complexity would be required for this purpose, and it must be a mechanism involving only known chemical and physical factors. It is safe to say that absolutely no hint as to the nature of this mechanism is contained in the hypothesis.

The determinants must be able to grow by reproduction, or by the accretion of new biophors, since in each generation new germ-cells are formed. If we say that they grow by reproduction in the sense that an organism grows by reproduction, we beg the question of their means of formation. Do they grow by the addition of similar substances in the way that a crystal grows? If so, the molecules of which they are composed must exist in the lymph stream bathing the germ-cells—that is, the biophors themselves must already exist in this liquid, for if we suppose that the biophors are able to divide and grow by making use of the protein substances which we know are present in the lymph stream, then we confer upon these bodies all the properties of the fully developed organism. If they are present in the blood, then the composition of the latter must be one of inconceivable complexity, since it must contain as many substances as there are distinct tissues in the animal body. We know, of course, that this is not the case. How, then, are the biophors reproduced?

We must leave this field of unbridled speculation (which cannot surely be " the most exact and rationalistic part of biology.") What the study of the reproduction of the organism does show is that something —which we call the specific organisation—is handed down from parent to offspring, and that this something *may* possess a high degree of stability. No apparent change of significance can be observed in the very numerous generation of organisms (the 2000 generations of *Paramœcium*, for instance, which were bred by Woodruff) which can be produced by experimental breeding. Some species of animals—the Brachiopod *Lingula*, for instance—have persisted unchanged since Palæozoic times. Throughout the incredibly numerous generations represented by this animal series, the specific organisation must have been transmitted in an almost absolutely unchanged condition. The germ-plasm is therefore continuous from generation to generation, and it possesses an exceedingly great degree of constancy of character. This conception of the continuity and stability of the specific organisation is the feature of value in Weismannism, and all that we know of the phenomena of heredity confirms it. But it is pure speculation to regard the organisation as an aggregate of chemically distinct substances, or if we say that this speculation is rather a working hypothesis, then it must justify itself by leading us back again to the results of experience.

It is, however, not quite accurate to say that the organisation persists unchanged from generation to generation. The offspring is similar to the parent— that is, the organisation has been transmitted unchanged. But the offspring also differs just a little from the parent—that is to say, the organisation is modified by each transmission. In these two statements we formulate in the simplest manner the law of organic variability. Organisms may obviously be arranged in categories in such a way that the individuals in any one category resemble each other more closely than they resemble the individuals belonging to another category. We may, by experimental breeding, produce an assemblage of organisms all of which have had a common ancestor, or a pair of ancestors. Now the individuals composing such an assemblage would exhibit a close resemblance to each other, such a resemblance as our categories of naturally occurring organisms are seen to exhibit. We should also find that the individuals of our naturally occurring assemblage would be able to interbreed among themselves, just as in the case of the experimentally produced population. It may be concluded, then, that the naturally occurring population is also the product of a pair of ancestors. This inter-fertility, as well as the close morphological resemblance of the individuals, are the facts on which the hypothesis of the common origin and unity of the assemblage, or species, is formed.

The morphological resemblance between the individuals, either in the natural or the artificial populations, is not absolute. If we take any single character capable of measurement we shall find that it is variable from organism to organism. This important concept of organic variability may be made more clear by a concrete example. Examination of a large number of cockle shells taken from the same restricted part of the sea-shore, and therefore belonging presumably to the same race, will show that the number of the radiating ridges on the shell varies from 19 to 27, and that the ratio of the length to the depth of the shell also varies from $1:0.59$ to $1:0.85$. In the former case the most common number of ridges is 23, and in the latter case the most common ratio of length to depth is $1:0.71$. These are the characteristic or modal values of the morphological characters in question, and the other or less commonly occurring values are distributed symmetrically on either side of the mean or modal value, forming " frequency distributions."[1] The value of the first character changes by unity in any distribution : obviously there cannot be a fraction of a ridge ; and this kind of variation is called " discontinuous." The value of the second character may change imperceptibly, and it is therefore called " continuous," a term which is not strictly accurate, since in applying it we assume that the numerical difference between two variates may be less than any finite number, however small. In this assumption we postulate for biology the distinctive mathematical concept of infinite divisibility.

The difference from the mode, or mean, with respect to a definite character in a fully grown organism may be due to the direct action of the environment, in the sense in which we have regarded the environment as influencing the organism ; or it may be due to the changes in the organism resulting from the increased or decreased use of some of its parts. The conditions with regard to nutrition, for instance, will not be the same for all the individuals composing a cluster of mussels growing on the sea-bottom. Those in the interior of the cluster do not receive so abundant a supply of sea-water as those on the outside of the cluster ; and since the amount of food received by any individual depends on the quantity of water streaming over it in unit time, we shall find that the internally situated individuals will be stunted or dwarfed, while those on the outside will be well grown. Such variations are acquired ones, but even when we allow for them, even if we take care that all the organisms studied live under conditions which are as nearly uniform as possible, there will still be some degree of variability. We cannot be sure that this absolute uniformity ever exists ; and the notion of

[1] See Appendix, p. 350.

the environment of an organism may be extended
so as to include the medium in which embryonic
development took place, and even the parental body
which formed the environment for the germ-cells
from which embryonic development began. But it
is probably the case that even with an uniform environment, or with one in which the differences
were insignificant, variability would still exist. The
variations that might be observed in such a case
would belong to two kinds—" fluctuating variations,"
and " mutations."

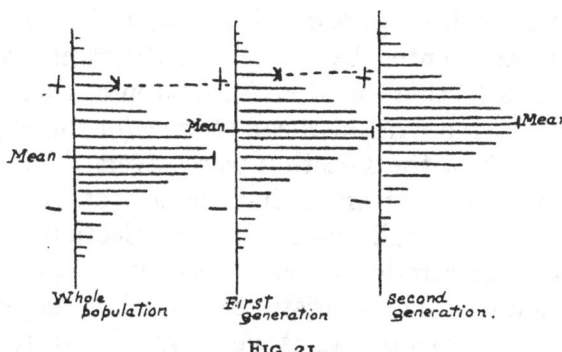

FIG. 21.

Whether the variations observed in a population
of organisms are fluctuations or mutations can only be
determined by experiment. Let us suppose that we
are dealing with a human population, and that the
variation studied is that of stature. Let the men
with statures considerably over the mean value marry
the women who are correspondingly tall, then it will
be found that the children from these unions will,
when grown up, exhibit a stature which is greater
than that of the whole population, but not so great
as that of their parents—that is, regression towards
the mean of the whole population takes place.

This is shown in the above diagram, where the

lines above and below the mean one indicate the proportion (relative to the value or frequency of the mean) of people of each grade of stature. The latter is proportional to the distance from the mean measured along the vertical line, distances below this line indicating statures below the mean, and *vice versa*.

If, on the other hand, the men and women with statures considerably below the mean marry, their children will ultimately exhibit statures which are greater than that of their parents, but which are less than that of the whole population. Regression again occurs, but in the opposite direction, and such a case would be represented by the above diagram reversed. Continued selection of this kind would lead to an immediate increase in the mean stature (or the opposite, if the " sign " of the selection were reversed) in one or two generations, but after that the amount of change would be very small, while if the selection were to cease the race produced would slowly revert to the mean, which is characteristic of the whole population from which it arose. It is very important to grasp this result of the practical and theoretical study of heredity—the selection of the ordinary variations shown by a general population leads at once to a small change in the mean value of the character which is selected, but continued selection thereafter makes very little difference to this result, while the race slowly reverts to the value of that from which it arose on the cessation of the selection.

Races which " breed true " do, of course, exist; thus the mean height of the Galloway peasant is greater than that of the Welsh. In the cases of " pure races "—that is, races which breed true with respect to one or more characters, we have to deal with another kind of variation, one which shows no tendency to revert to the value from which it arose. Let the observed variability of stature in a human population be represented by the frequency distribution A, and let the individuals at N—that is, those in which the stature was greater than the mean by the deviation ON—intermarry. It might then happen that the variability of the offspring of these unions would be represented by the frequency distribution B, in which the value of the mean is also that of the stock, at N, from which the race originated. It does

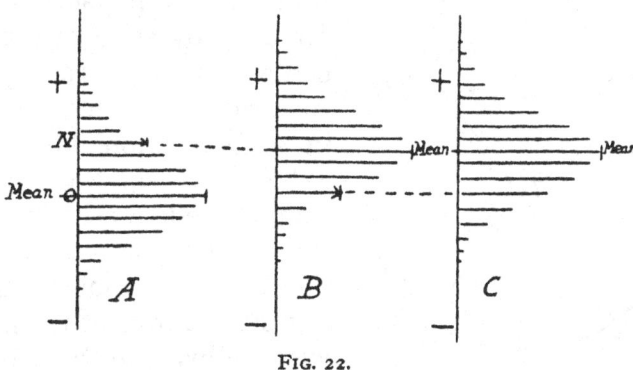

FIG. 22.

not matter now from what variants in B a progeny of the third generation arises : the mean height of the latter will be that of the pure race.   In this case the individuals from which the pure race originated (those at N in A) have exhibited a mutation. The stature of the individuals of this new race will continue to exhibit fluctuating variations, and the range of this variability may be as much as that of the stock from which it arose, *but the mean stature of the new race* will continue to be that of the original mutants.

It is well known that de Vries himself considered fluctuating variations and mutations as something quite different.   The former he considered as nothing new, only as augmentations or diminutions of something previously existing ; and he regarded fluctuations as due to the action of the environment, following in their distribution the laws of chance.[1]   Mutations, on the other hand, were something quite new.   Now future analysis of variability will not, we think, bear out the validity of this distinction.   It is far more likely that a fluctuation is a variation which is the result of some causes the action of which is variable. (We are regarding variability now as subject to " causation " in the physical sense, for only by so regarding it can we attempt its analysis). As a rule this process results in a fluctuation, but if its extent, or degree of operation, exceeds a certain " critical value " a mutation is produced.   We may, following the example of the physicists, illustrate this by a " model."

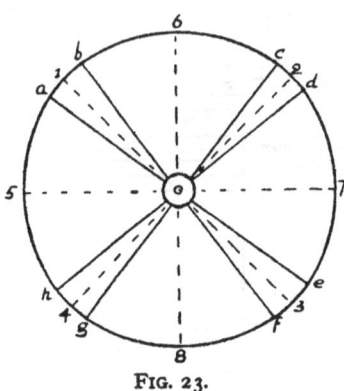

FIG. 23.

This model is a modification of Galton's illustration of the degrees of stability of a species.   It is a disc of wood rolling on its periphery.   We divide it into sectors, and the arcs *ab, cd, ef,* and *gh* have all the same radius, 10, 20, 30, and 40.   Then we flatten the sectors *bc, de, fg,* and *ha,* so that their radii are greater than are those of the other arcs.   Now let us cause the disc to roll about the point 8 as a centre.   It will oscillate backwards and forwards about a mean position 8.   Let us think of these oscillations as fluctuations.

Suppose, however, that we cause the disc to roll

<hr>

[1] See Appendix, p. 351.

a little more violently, so that it oscillates until either of the points 3 or 4 are perpendicularly beneath the centre O. In either of these positions the disc is in a condition of " unstable equilibrium," and an infinitesimal increase in the extent of an oscillation will cause it to roll beyond the points 3 or 4. But if it does pass either of these critical points it will begin to oscillate about either of the new centres 5 or 7, thus rolling on one of the arcs, *ha* or *de*. This assumption of a new condition of stability we may compare with the formation of a mutation.

All this is merely a conceptual physical model of a process about which we know nothing at all. It is meant to illustrate the view that the organisation of a plant or animal is not something absolutely fixed and invariable. The organism in respect of each recognisable and measurable character oscillates about a point of stability, that is to say exhibits fluctuating variations about the mean value of this character. If the stability of the organisation is upset, so that it oscillates, or fluctuates about a new centre, that is, if the variations deviate in either direction from a new " type " or mean, a mutation has been established. A mutation is not, therefore, necessarily a large departure from " normality." It is not necessarily a " discontinuous variation," nor a " sport " nor a " freak." It is essentially a shifting of the mean position about which the variations exhibited by the organism fluctuate.

Such a mutation will, in general, involve the creation of an " elementary species." We have considered only one character, say stature, in the above discussion, but it generally happens that the assumption of a new centre of stability involves *all* the characters of the mutating organism. An elementary

N

species therefore differs a little in respect of all its characters from the species from which it arose, or from the other elementary species near which it is situated. This is what we do usually find in the cases of the " races," or " local varieties," of any one common species of plant or animal. That we do not recognise that most, or perhaps all, of the species known to systematic biology are really composed of such local races is merely because such results involve an amount of close investigation such as has not generally been possible except in the few cases studied with the object of proving such variability ; or in the case of those species which are studied with great attention to detail because of their economic importance. Thus the herrings of North European seas can be divided into such races, and it is possible for a person possessing great familiarity with these fishes to identify the various races or elementary species—that is, to name the locality from which the fish were taken—by considering the characteristics in respect of which the herrings of one part of the sea differ from those of other parts.

The term " variety " has rather a different connotation in systematic biology from that which is included by the term " elementary species." The meaning of the latter is simple and clear. Two or more elementary species are assemblages of organisms, in each of which assemblages the mean positions about which the various characters fluctuate is different. The term "variety" cannot so easily be defined. The progeny of two different species (in the sense of the term as it is usually applied by systematists) may be called a hybrid variety of one or other of the parent species. In the case of the ordinary species of zoology such a hybrid would, in general, be infertile, or if it did produce offspring these would be infertile.

In the case of ordinarily bred offspring from parents of the same species a large deviation from the parental characters might be a malformation, or the result of some irregularity of development. An " atavistic " variation we may regard as the reappearance of some character present in a more or less remote ancestor. Thus dogfishes and skates are no doubt descended from some elasmobranch fish which possessed an anterior dorsal fin. This fin persists in the dog-fishes, but has been lost in the skates and rays. Yet it may appear in the latter fishes as an atavistic variation.

In a variety (following de Vries' analysis) a character which disappears is not really lost : it is only suppressed, and it still exists in a latent form. Some flowers are coloured, for instance, but there may be varieties in the species to which they belong in which the flowers are colourless. It may not be quite correct, in the physical sense, to say that the colour has been lost, but we may put it in this way. These flowers are then coloured and colourless varieties of the same species. Colour or lack of colour is not, however, fixed in the variety, for the individual plant bearing colourless flowers also bears in its organisation the potentiality of producing coloured flowers. The petals of a flower may be smooth or covered with hairs, and in the same stock both of these varieties may occur. But we must not speak of the presence or absence of hairs as constituting a difference of kind : the smooth-petalled flowers might be regarded as containing the epidermal rudiments of hairs. So also coloured and colourless flowers may be regarded as containing the same kinds of pigment, but these pigments are mixed in different proportions. Such a view enables us to look upon these contrasting characters in the same way as we look upon fluctuating variations, that is, as quantitative differences in the value of the same character.

Such a suppression of a character is not really a loss. An organism belonging to an elementary species in which, say, monochromatic flowers are usually produced may produce flowers which are striped. The progeny of the plant may still produce monochromatic flowers, but we must think of it as also possessing the potentiality of producing striped flowers. In the terminology of Mendelism the characters are dominant and recessive ones.

In discussing Mendelian varieties we consider the manner in which two contrasting characters—one present in the male parent and one in the female— are transmitted to the offspring. The characters in question may be the tallness of the male parent and the contrasting shortness of the female ; or the brown eyes of the male and the blue eyes of the female ; or the brown skin of the female parent and the white skin of the male one. These characters may be inherited in two ways : either they may be blended or they may remain distinct in the offspring. The children of the brown mother and the white father are usually coloured in some tint intermediate between those of the parents. The mulatto hybrid is fertile with either of the parent races, and again the offspring may take a tint intermediate between those of the parents, and so on through a number of generations. But somewhere in this series the concealed or recessive brown colour may appear in all its completeness, showing that it has been present in the organisations of all the intervening generations. The progeny of a tall male parent and a short female parent are not, in general, intermediate in stature between the parents ; some of them may be tall and others short. The children of a brown-eyed mother and a blue-eyed father do not usually have eyes in which the colours of the parental eyes are blended : they are blue-eyed or brown-eyed. The contrasting characters are spoken of as dominant and recessive : if tallness is transmitted to offspring, which may nevertheless produce dwarf offspring, the latter character is said to be recessive to tallness. The contrasting characters of the parents therefore remain distinct in the progeny, some of the latter exhibiting the one character and some the other ; while it may happen that the one character or the other may be segregated, so that it only appears in, and is transmitted by, the offspring. There are numerical relationships between the numbers of the offspring in which the contrasting characters appear.

Obviously, tallness and dwarfness are not characters which differ in *quality :* they are different degrees of the same thing. Brown eyes and blue eyes are not necessarily different in quality, for we may think of the same kinds of pigment as being present in the iris but mixed in different proportions. But the terminology of this branch of biology appears to suggest that the contrasting characters are, each of them, something quite different from the other : there are "factors" for "tallness," "dwarfness," for blue eyes and brown eyes, and so on. These qualities are called "unit-characters," and they are supposed to possess much the same individuality in the germ-plasm as the "radicles" of the chemist possess in a compound. Sodium chloride, for instance, is not a blend of sodium and chlorine : the two kinds of atoms do not fuse together but are held together merely. The analogy is, however, very imperfect, for in the chemical molecule the characters are not those of either of the constituents but something quite different, whereas in the Mendelian cross the characters remain distinct, but one of them is patent while the other is latent. In the molecule, however, the atoms are regarded by the chemist as lying beside each other in certain positions, and the Mendelian factors are also spoken of as if they lay side by side in the germ-plasm. This terminology is useful, perhaps necessary, in the work of investigation, but we must not forget that it symbolises, rather than describes, the results of experiment. If the factors are identified with certain morphological structures in the nuclei of the germ-cells, obviously all the objections that may be urged against the Weismannian hypothesis as an hypothesis of development apply also to the Mendelian hypotheses as descriptions of a physical process of the transmission of morphological characters.

It should clearly be understood what is implied in the construction of such a hypothesis. Certain processes are observed to take place when a somatic cell divides : these processes we have regarded as having for their object the exact division of all the parts of the cell into two halves. This process of somatic cell division is modified when a germ cell divides prior to maturation (the process fitting it to become fertilised). Then the cell nucleus divides into four daughter-nuclei. One of these remains in the cell substance which is to become the ovum, and the other three, each of them invested in a minimal quantity of cytoplasm, are eliminated as the " polar bodies." Also the number of chromosomes in the mother-cell becomes halved, so that the mature ovum, or spermatozoon, possesses only one-half of the number of chromosomes which are present in the ordinary somatic cell. Now let the reader puzzle out for himself what may be meant by this behaviour of the germ cells, and he will certainly see that several interpretations are possible. But suppose that the chromatin consists of an incredibly large number of bodies differing in chemical structure from each other, and occupying definite positions with regard to each other ; and suppose that there is a mechanism of unimaginable complexity in the cell capable of rejecting some of these chemically individualised parts, and of " assembling " or arranging the others in much the same way as an engineer assembles the parts of a dynamo when he completes the machine. *Then* we may regard the hypothetical discrete bodies which form the hypothetical nuclear architecture as the material carriers of Mendelian characters. It is strange that the correspondence of such a logically constructed mechanism with the effects which it would produce if it existed should be regarded as a proof that it does exist, yet biological speculation has actually made use of such an argument. " It seems exceedingly unlikely that a mechanism so exactly adapted to bring it " (the separation from each other of the Mendelian material " factors " of inheritance) " about should be found in every developing germ cell if it had no connection with the segregation of characters that is observed in experimental breeding." Put quite plainly this argument is as follows : there is a certain segregation to be seen in experimental breeding, and certain processes may be observed to occur in the developing germ cell. Add to these processes many others logically conceivable, and add to the observed material structure of the cell another structure also logically conceivable. Then the assumed mechanism and structure is " exactly adapted " to produce the effects which are to be explained.

Therefore the mechanism and structure do actually exist !

That which renders the son similar to the father—the specific organisation—is undoubtedly very stable, and it may persist in the face of a variable environment. But now and then the son differs from the father. The differences may be " accidental " and may not be transmitted further—then we have to deal with an unstable fluctuation ; or the differences may be permanent—then we have to deal with a stable mutation. What " produces " a mutation ? A change of the environment, it may be said : if so, the mutation is an active change or adaptation of the organism to a change in its surroundings, and this adaptation is a permanent one and is transmitted. Or the mutation may be a spontaneous change of functioning. If this disturbance of the stability of the organisation is *general*, if it affects all the characters of the organism, we have to deal with the establishment of a new elementary species. But if the disturbance affects only one, or a few characters, then we need not recognise that a new elementary species has come into existence. Men and women remain men and women (in their morphology), although some time or other among the brown eyes characteristic of a race blue eyes may have appeared. The result of the disturbance, in this case, has been to cause one, or a few, of the characters that fluctuate to surpass their limits of stability.

The idea of the elementary species is a clear and simple one. It is a group of organisms connected by ties of blood relationship : all have descended from one pair of ancestors. The individuals exhibit certain characters, all of which are variable. This variability is not cumulative ; in generation after generation the individuals of the species display variations which fluctuate round the same mean values. Two or more elementary species may have had the same origin— a common ancestor or ancestors—but the organisms in one species exhibit characters which, although similar in nature to those of the other species, yet fluctuate about different mean values.

This is not the " species " of the systematic biologist. The Linnean or systematic species is a concept which is much more difficult to define : it is a concept indeed which has not any clear and definite meaning, in actual practice.

We often forget how very young the science of systematic biology is, and how intimately its progress has been dependent on that of human invention and industrial enterprise. Physics and mathematics might be studied in a monastic cell, but the study of systematic biology can only be carried on when we have ships and other means of travelling—the means, in short, of collecting the animals and plants inhabiting all the parts of the earth's surface. Until a comparatively few years ago the fauna and flora of great tracts of land and sea were almost unknown : even now our knowledge of the life of many parts of the earth is scanty and inaccurate. Systematic biology has therefore had to collect and describe the organisms of the earth, and in so doing it has set up the Linnean species of plants and animals. These we may describe as, in the main, categories of morphological structures. The older and more familiar species are clearly defined in this respect : such are cats and dogs, rabbits, tigers, herrings, lobsters, oysters, and so on : the individuals in each of these categories are clearly marked out with respect to their morphology, and the limits of the categories are clearly defined. In all of them the specific organisation has attained a high degree of stability so that the individuals " breed true to type " ; and it has also attained a high degree of specialisation, so that it does not fuse with other organisations.

Yet, in the marjoity of the systematic species of biology, this criterion of specific individuality—this recognition of the isolation of the species from other species—cannot be applied. Very many species have been described from a few specimens only, many from only one. How does a systematist recognise that an organism with which he is dealing has not already been classified ? It differs from all other organisms most like it, that is, he cannot identify it with any known specific description. But the differences may be very small, and if he had a number of specimens of the species most nearly resembling it he might find that these differences were less than the limits of variation in this most closely allied species, and he would then relegate it to this category. But if he has to compare his specimen with the " type " one, that is, the only existing specimen on which the species of comparison was founded, the test would be unavailable. The question to be answered is this : are the difference or differences to be regarded as fluctuations, or are they of " specific rank " ? Now certainly many systematists of great experience possess this power of judgment, though they might be embarrassed by having to state clearly what were the grounds on which their judgment was based. But on the other hand hosts of species have been made by workers who did not possess this quality of judgment ; and even with the great systematists of biology confusion has originated. Slowly, very slowly, the organic world is becoming better known, and this confusion is disappearing.

The species, then, whether it is the systematic group of the biological systems, or the elementary species based on the study of variability and inheritance, is an intellectual construction : an artifice designed to facilitate our description of nature. This is clearly the case with the higher orders of groups in classifications : genera, families, orders, classes, and phyla express logical relationships, or describe in a hypothetical form our notions of an evolutionary process. But species, it may be said, have an actual reality : there are no genera in nature, only species. These categories of organisms really exist ; they have individuality, a certain kind of organic unity, inasmuch as the individuals composing them have descended from a common ancestor. Yet just as much may be said of genera, families, and the other groupings. One species originates from another by a process of transmutation : a genus is a group of species which have all had a common origin ; a family is a similarly related group of genera, and so on. The higher categories of biological science are intended to introduce order and simplification into the confusion and richness of nature as we observe it, but obviously the concept of the species has the same practical object. Must we then say that there are no species in nature, only individuals ? If so, we are at once embarrassed by the difficulty of forming a clear notion of what is meant by organic individuality. Does it not indicate that life on the earth is really integral, and that our analysis of its forms—species, genera, families, and so on—are only convenient ways of dealing actively with all its richness ?

Systematic biology is a very matter-of-fact occupation, and one is surprised to find upon reflection how he, in his handling of the concepts of the science, follows the methods of ancient philosophy. In classical metaphysical systems mutability was an illusion. Behind the confusion and change given to sensation there is something that is immutable and eternal. If there is change there is something that changes ; or, at least there ought to be something that changes when it is perceived through the mists of sensation, just as the image of a well-known object on the horizon wavers and is distorted by refraction. This immutable reality is the Form or Essence of the Platonic Idea : that which is in some way degraded by its projection into materiality, so that we become aware of it only through our imperfect organs of sense. We do not see the Form itself, but its quality rather, the Form with something added or something taken away from it.

The Form itself is only a phase in a process of transmutation. Everything that exists in time flows or passes into something else. But it is not a momentary or instantaneous view of the flux that we see, but rather a certain aspect of the reality that flows, that in some way expresses the nature of the transmutation from one Form into another. The sculptor represents the motion of a man running by symbolising in one attitude all the actions of body and limbs ; so that from our actual, sensible experience or intuition of the movement of the runner we see in the rigid marble all the plasticity of life. The instantaneous photograph shows us a momentary fixed attitude of the runner—an attitude which is strange and unfamiliar. The Idea does not, then, represent a moment of becoming like the photograph, but rather a typical or essential phase of the process of transmutation, just as the sculptor represents in immobile form the characteristic leap forward of the runner. Just as our intuitive knowledge of the actions of our own bodies enables us to read into the characteristic attitude represented in the marble all the other attitudes of the series of movements, so our experience enables us to expand the formal moment of becoming into the action which it symbolises.

This action has a purpose, an intention or design which was contemplated before it began. There is therefore the threefold meaning in the Platonic Idea : (1) an immutable and essential Form of which we perceive only the quality ; (2) the characteristic phase in the transmutation of this Form into some other one ; and (3) the design or intention of the transmutation.

This was, as Bergson says, the natural metaphysics of the intellect. It was, in reality, the " practical " way of introducing order and simplification into the confusion of the sensible world—all that is presented to us by our intuitions. And in the effort to reduce to order the welter of the organic world biology has followed the same method, so that it represents the species with the threefold significance of the Platonic Idea. That which is expressed in the term species is an assemblage of organisms each of which is defined by an essential form and an essential mode of behaviour —the characters indicated in the specific diagnosis. But organisms are variable, their specific characters fluctuate round a mean, and in saying this we suggest that there is something which varies—there *ought to be* an essential form from which the observed forms of the individuals deviate, something invariable which nevertheless varies accidentally. This is (1) the quality of the specific idea. So also we never do actually observe the essential individual ; what we do see is the embryo, or the young and sexually immature organism, or the sexually mature one, or the senescent one : there is continual change from the time of birth to that of senile decay. This confusion is unmanageable, and for it we substitute the characteristic form and functioning, and that phase in the life-history of the organism which suggests all that the previous phases have led up to, and all that subsequent phases take away. Thus there is contained in our idea of the species (2) the notion of a typical moment in an individual transformation. It is not a " snap-shot " of some moment in the life-history that we make : in identifying a larval form as some species of animal we are identifying it with all the other phases of the life-history.

Since we accept the doctrine of transformism, the specific idea also includes that of an evolutionary process. For the organic world is a flux of becoming, and species are only moments in this becoming. It does not help us to reflect that if the hypothesis of evolution by mutations is true the process is a discontinuous one : mutability is the result of periods of immutability during which the change was germinating, so to speak. In this flux of becoming we seize moments at which the specific form flashes out—not as instantaneous views of the flux, but as aspects of it which suggest the steps, the morphological processes, by which the transmutation of the species has been effected. Thus our specific idea represents not only a phase of becoming in an individual life-history, but also a phase of becoming in an evolutionary history.

Whether we consider this evolutionary movement as the working out of a Creative Thought, or as the development of elements assembled together by design, or as the results of the action of a mechanism working by itself, we must suppose that underlying it there is design, or purpose, or determinism. All is given, therefore, and our comparison between the metaphysical Platonic Idea and the modern concept of the species becomes complete.

CHAPTER VI

THE species is therefore a group of organisms all of which exhibit the same morphological characters. This sameness is not absolute, for the individuals composing the species may vary from each other with respect to any one character. But the range of these variations is limited. They fluctuate about an imaginary mean value which remains constant in the case of a species which is not undergoing selection, and is therefore nearly the same throughout a series of generations. The formal characters which we regard as diagnostic of the species are these imaginary mean ones.

It is possible to breed from stock a very great number of animals, all of which are connected by a tie of blood-relationship, that is, all have descended from the same ancestor or ancestors. Such an assemblage of animals would resemble those assemblages living in the wild which we call species, in that a certain morphological similarity would be exhibited by all the individuals. If the breeding were conducted so as to avoid selection, the range of variability would be very much the same as that observed in the wild race. The two groups of animals—that bred artificially, and that observed in natural conditions—would be very much alike, and it is impossible to resist the conclusion that the natural race, like the artificial one, is a family in

the human sense, that is, all the individuals composing it are connected together by a tie of common
descent.

Let us extend this reasoning to categories of organisms of higher orders than species. We can associate
together groups of species in the same way that we
associate together the individuals of the same species.
There are certain morphological characters which are
common to all the species in the category, but there
are also differences between specific group and specific
group, and these differences may be regarded as
variations from the generic morphological type. All
the Cats, for instance, have certain characters in
common : fully retractile claws, a certain kind of
dentition, certain cranial characters, and so on. We
postulate a feline type of structure, and we then
regard the characters displayed by the cat, lion, tiger,
leopard, etc., as deviations from this feline morphological type. Thus we establish the Family Felidæ.
But again we find that the Felidæ together with the
Canidæ, and many other species of animals, also display
common characters, dental and osteological chiefly,
and we express this resemblance by assembling all
these families in one Order, the Carnivora. The
Carnivores, however, are only one large group of
Quadrupeds : there are many others, such as the
Rodents, Ungulates, Cetacea, etc., and all of these
possess common characters. In all of them the
integument is provided with hairs, or other similarly
developed structures ; all breathe by means of a diaphragm ; in all, the young are nourished by suckling
the mammæ of the mother ; and all develop on a
placenta. We therefore group them all in the Class
Mammalia. Now the Mammals possess an internal
skeleton of which the most fundamental part is an

o

axial rod—the notochord—developing to form a vertebral column ; and this notochordal skeleton is also possessed by the Birds, Reptiles, Amphibia, and Fishes. There are also some smaller groups in which the notochord is present but does not develop to form segmented vertebræ. Including these, we are able to form a large category of animals—the Chordata— and this phylum is sharply distinguished from all other cognate groups.

All animals and plants may be classified in a similar way. Insects, Spiders, and Crustacea, for instance, are all animals in which the body is jointed, each joint or segment being typically provided with a pair of jointed appendages or limbs. Because of this similarity of fundamental structure we include all these animals, with some others, in one phylum, the Arthropoda. So also with the rest of the animal kingdom, and similar methods may be extended to the classification of the plants. A few small groups in each of the kingdoms are difficult to classify, but it has been possible to arrange most living organisms in a small number of sub-kingdoms or phyla, and even to attempt to trace relationships between these various categories.

The mere systematic description of the organic world would have resulted in such a reasoned classification apart altogether from any notions of an evolutionary process. But the classification, originally a conventional way of making a list of organisms, would at once suggest morphological similarities. It would suggest that all the Cats were Carnivores, that all the Carnivores were Mammals, and that all the Mammals were Chordates. It would suggest that all Wasps were Hymenoptera, that all Hymenoptera were Insects, and that all Insects were Arthropods. It would establish a host of *logical* relations between animals of all kinds.

It would show us a number of groups of animals separated from each other by morphological dissimilarities. But let us also consider all those animals which lived in the past of the earth, and the remains of which are found in the rocks as fossils. Including all the forms of life known to Palæontology, we should find that the dissimilarities between the various groups would tend to disappear. The gaps between existing Birds and Reptiles, for instance, would become partially bridged. Palæontology would also supplement morphology in another way. The study of the structure of animals leads us to describe them as " higher " and "lower "— higher in the sense of a greater complexity of structure. Thus the body of a Carnivore is more complex than that of a Fish, inasmuch as it possesses the homologues of the truly piscine gills, but it also possesses a four-chambered heart instead of a two-chambered one ; and it possesses the mammalian lungs, diaphragm, and placenta, structures which are not present in the Fish. Now, so far as its imperfect materials go, palæontology shows us that the higher forms of life appeared on the earth at a later date than did the lower forms. The remains of Mammals, for instance, are first found in rocks which are younger than (that is, they are superposed upon) those rocks in which Reptiles first appear ; and so also Reptiles appear later in the rock series than do Fishes. Palæontology thus adds to the logical order suggested by morphology a *chronological* order of this nature : higher, or more complex forms of life appeared at a later date in the history of the earth than did lower or less complex ones.

A parallel chronological sequence would also be suggested by the results of embryology. This branch of biology shows us that all animals pass through a series of stages in their individual development, or ontogeny. The earlier stages represent a simple type of structure, usually a hollow ball of cells, but as development proceeds, the structure of the embryo becomes more and more complex. The process of development is continuous in many animals, but in others (perhaps in most) larval stages appear, that is, development is interrupted, and the animal may lead for a time an independent existence similar to that of the fully developed form. Often these larval stages suggest types of structure lower than that of the fully developed animal into which they transform. Even if larval stages may not appear in the ontogeny, it is very often the case that the developing embryo exhibits traces, or at least reminiscences, of the types of morphology characteristic of the animals which are lower or less complex than itself ; thus the piscine gills appear during the development of the tailed Amphibian, and even in that of the Mammal, and then vanish, or are converted into organs of another kind. The individual thus passes through a series of developmental stages of increasing complexity : it repeats, in its ontogeny, the palæontological sequence in a distorted and abbreviated form.

It is true that the evidence afforded by palæontology is very meagre. The preservation of the remains of organisms in the stratified rocks is a very haphazard process, and it depends for its success on a series of conditions that are not always present. As the surface of the earth becomes better known, our knowledge of the life of the past will become fuller, but there can be little doubt that whole series of organisms must have existed in the past, and that no recognisable traces of these are known to us. There is also no doubt that the sequences indicated by palæontology are very incomplete : they are obscured and shortened by many conditions. The earlier embryologists entertained hopes that the study of embryology would reveal the direction of the evolutionary process in many groups of animals : if the organism repeats in its ontogeny the series of stages through which it passed in its phylogenetic development, then a close study of the embryological process ought to disclose these stages. Although these hopes have not been realised, there is yet sufficient truth in the doctrine of recapitulation to enable us to state that there is a rough parallelism between the palæontological and embryological sequences.

We therefore state a plausible hypothesis when we assert that different species may be related to each other in the same way that the individuals of the same species are related, that is, by a tie of blood-relationship ; and that different genera, families, orders, and so on are also so related. Morphological studies enable us to arrange numbers of species in such a way that series, in each of which there is an increasing specialisation of structure, are formed. Both palæontology and embryology show, to some extent at least, that these stages of ever-increasing specialisation of structure occurred one after the other. Now, stated briefly and baldly as we have put it, this argument may not appear to the general reader to possess much force, but it is almost impossible to over-state the strength of the appeal which it makes to the student of biology. To such a one a belief in a process of transformism will appear to be inseparable from a reasoned description of the facts of the science.

But it would be no more than a belief, not even a hypothesis, if we did not attempt to verify it experimentally. It is merely logical relationships that we establish, and the chronological succession of forms of life, higher forms succeeding lower ones, does not itself do more than suggest an evolutionary process. All that we have said is compatible with a belief in a process of special creation. But if we cling to such a belief, if we suppose that the organisms inhabiting the earth, now and in the past, are the manifestations of a Creative Thought, we must still accept the notion of logical and chronological relationships between all these forms of life. If we permit ourselves to speculate on the working of the Creative Thought, we seem to recognise that the ideas of the different species must have generated each other, and that the genesis of living things must have occurred in some such order as is indicated by a scientific hypothesis of transformism. An evolutionary process must have occurred somewhere, but the kinships so established between organisms would be logical and not material ones.

Science must not, of course, describe the mode of origin of species in this way. So long as it investigates living things by the same methods which it uses in the investigation of inorganic things, it must hold that the concepts of physical science are also adequate for the description of organic nature. It must assume that matter and energy and natural law are given ; and that, even in the conditions of our world, life must have originated from lifeless matter ; must have shaped itself, and undergone the transformations that are suggested by the results of biology. It must assume, in spite of the formidable difficulties that the assumption encounters, that cosmic physical processes are reversible and cyclical ; and that worlds and solar systems are born, evolve, and decay again. Every stage in such a cosmic process, as well as every stage in the evolution of living things, must have been inevitably determined by the stages preceding it. Such a mechanistic explanation must assume that a superhuman intellect, but still a finite intellect like our own, such a calculator as that imagined by Laplace or Du Bois-Reymond, would be able to deduce any state of the world, or universal system, from any other state, by means of an immense system of differential equations. It would be able, as Huxley says, to calculate the fauna of Great Britain from a knowledge of the properties of the primitive nebulosity with as much certainty as we can say what will be the fate of a man's breath on a frosty day. Such a fine notion as that of an universal mathematics must ever remain as the ideal towards which science strives to approximate.

Or we may suppose that a plan or design has been superposed on nature, is immanent in matter and energy, and works itself out, so to speak. Such a teleological explanation of inorganic and organic evolution inevitably forces itself upon us if we reject the notion of radical mechanism. We think of an universal system of matter and energies as consisting of elements which, when assembled together, interact in a certain way, and with results which are definite and calculable. The assembling together of the elements of the system would be the result of the previous phases of the system. That is radical mechanism. But let us think of the elements of the system as being differently assembled—thus involving the idea of an agency, external to the system, which rearranges them—then the same energies inherent in this system, as in that previously imagined, will also work out by themselves. But the result will be different, and will depend on the manner in which the elements were originally arranged. That would be radical finalism.

Science must reject this notion as it rejects that of special creation, since it introduces indeterminism into the evolutionary process. It must regard the organism and its environment as a physico-chemical system studied from without. It must avoid all attempts to acquire an intuitive knowledge of the actions of the organism, for the latter, and the things which environ it, are only bodies moving in nature. In the systems studied by it time must be the independent variable, and there must be a strict functionality between the parts of the organism and the parts of the reacting environment, so that any change in the one must necessarily be dependent on a change in the other. Such a system and series of interactions is that which is described in a mechanistic hypothesis of transformism.

All this is indeed suggested to ordinary and aided methods of observation. The plant or animal acts upon, and is acted on by, the environment, though it is usually the modification of the organism to which we attend. A man's face becomes reddened by wind and sun and rain ; manual labour roughens his hands and develops callosities ; in the summer he sweats and loses heat ; in the winter the blood-vessels of his skin contract and heat is economised. In the winter months the fur of many animals becomes more luxuriant and may change in colour. Fishes which inhabit lightly coloured sand are lightly pigmented, but their skins become dark when they move on to darkly coloured sea-bottoms ; prawns which are brown when they live on brown weed, become green when they are placed on green weed. Birds migrate into warmer countries, and *vice versa*, when the seasons change.

Such are instances of the adaptations of the morphology and functioning of organisms consequent on
changes of environment.

What is an adaptation ? The term plays a great
part in biological speculation, but it is often used in a
loose and inaccurate manner, and not always in the
same sense. It suggests that the organism is *contained*
by the environment, and that its form becomes
adapted to that of the latter, just as the metal which
the ironfounder pours into the mould takes the form
of the cavity in the sand. " We see once more how
plastic is the organism in the grasp of its environment "—such a quotation from morphological literature is perhaps a typical one. Over and over again
this passive change in the organism as the result of
the action of something rigid which presses upon it
is what is understood by an adaptation. No doubt
the organism may be so affected, and often the change
which it experiences is of the same order as the environmental change. In the winter many animals
become sluggish and may hibernate ; their heart-beats
slow down ; their respiratory movements become less
frequent, and generally the rate of metabolism, that
is the rapidity with which chemical reactions proceed
in their tissues, becomes lessened. All these changes
become reversed in sign when the temperature again
rises. The time of year at which a fish spawns depends on the nature of the previous season. The rate
of development of the egg of a cold-blooded animal
varies with the temperature. The quantity of starch
formed in a green leaf depends on certain variables
—the intensity of light, the temperature, and the
quantity of carbonic acid contained in the medium
in which it is placed. In all these cases the rate at
which certain metabolic processes go on in the body

of an organism varies according to the conditions of the environment. In general they are cases of van't Hoff's law, that is, the rapidity at which a chemical reaction proceeds varies according to the temperature.

They are changes of functioning passively experienced by the organism as the result of environmental changes, and we must clearly distinguish between them and such changes as are the result of some activity or effort on the part of the organism. A flounder which lives in a river migrates out to sea when the first of the winter snows melt and flood the estuary with ice-cold water. Brown or striped prawns living on brown or striped weeds become green when they are placed on green weed, changing their pigmentation to match that of the alga. A kitten brought up in a cold-storage warehouse develops a sleeker and more luxuriant coat than does its sister reared in a well-warmed house. An animal which recovers from diphtheria forms an antitoxin which enables it to resist, for a time at least, repeated infection. A man who goes exploring in polar seas puts on warmer clothing than he wears in the tropics.

It is not necessary that an environmental change should occur in order that an adaptation should be evoked, for the organism may react actively and purposefully to a change in itself. The athlete acquires by running or rowing a more powerful heart ; the blacksmith develops more muscular shoulders and arms ; and the professional pianist more supple wrists and fingers. If one kidney is removed by operation, or if one lung becomes diseased, the organ on the other side of the body becomes hypertrophied. Aphasia, which is due to a lesion in the unilateral speech-centre, may pass away if the previously unused centre on the other side of the brain should become functionally active. In general, the continued use of an organ leads to its increase in size and efficiency, and conversely disuse leads to a decrease of size and even to atrophy.

The essence of an adaptation is that it is an active, purposeful change of behaviour, or functioning, or morphology, by which the organism *responds* to some change in its physical environment, or to some other change in its own behaviour, or functioning, or morphology. It is also a change which remains as a permanent character in the organisation of the animal exhibiting it. It does not matter even if the change of behaviour is one which is willed in response to some change of environment actually experienced, or whether it anticipates some change that is foreseen. A changed mode of behaviour adapted intelligently leaves, at the least, a memory which becomes a permanent part of the consciousness of the animal, and may influence its future actions ; or if it is evoked by a process of education it must involve the establishment of a " motor habit." The education of a singer sets up, in the cortex and lower centres of the brain, a nervous mechanism which controls and co-ordinates the muscles of the chest and larynx, and which did not exist prior to the process of education. Adaptations are therefore acquired changes of some kind or other by means of which the organism is able to exert a greater degree of mastery over its environment, including in the latter both the inert matter of inorganic nature and the other organisms with which the animal competes.

They are acquirements because of which the organism deviates from the morphological structure characteristic of the species to which it belongs. Do they affect the entire organisation of the animal exhibiting them, that is, may an acquired change of structure be so fundamental that it affects not only the body of the animal in which it occurs but also the progeny of this animal ? Let us suppose that this is the case ; let us suppose that quite a large proportion of all the individuals of a species inhabiting a restricted part of the earth's surface acquire the same change of character simultaneously ; and that they transmit this deviation of structure to their progeny. Then we should have an adequate means whereby the specific type becomes modified—a means of transformism.

This is the hypothesis which is associated with the name of Lamarck, and its essential postulate is that characters which are acquired by an organism during its own lifetime are transmitted to its offspring. It seems reasonable to suppose that this transmission of acquired characters should occur—how reasonable we should note when we see that de Vries tacitly assumes that fluctuating variations due to the action of the environment may be inherited by the offspring of organisms which exhibit them. That transmutation of species might occur in this way was a popular and widespread belief in England and Germany throughout the greater part of the nineteenth century ; and it was a belief entertained by Darwin himself, and confidently, and even dogmatically affirmed at one time by the majority of biologists in both countries.

How was it, then, that a very general change of opinion with regard to this question occurred both in England and Germany during the last two decades of the last century ? Certainly many botanists and zoologists continued to adhere to the older hypothesis, and most physiologists still do not appear to make any clear distinction between morphological characters which are inherited and those which are acquired ; but the majority of biologists did not hesitate to conclude that not only was the transmission of acquired characters an unproved conjecture, but that it was even theoretically inconceivable. At the beginning of the nineteenth century this belief had almost become a doctrine dogmatically asserted, and one cannot fail to notice a tone of irritation and impatience on the part of the spokesmen of zoology when the contrary opinions are expressed. " Nature," says Sir E. Ray Lankester, " (and there's an end of it) does not use acquired characters in the making and sustaining of species for the very simple reason that she cannot do so."

There can be little doubt that the interrogation of nature with regard to this question was not a very thorough process. The dogmatic denial of the transmission of acquired characters was not the result of exhaustive experiment and observation, but was due rather to the very general acceptance in England and Germany of Darwin's hypothesis of the transmutation of species by means of natural selection, and of Weismann's hypothesis of the continuity of the germ-plasm.

The newer hypothesis of transmutation was one which seemed adequate to account for the diversity of forms of life, so that it was unnecessary to invoke the older one ; though Darwin himself admitted that the individual acquirement of structural modifications might be a factor in the evolutionary process ; and for more than twenty years after the publication of the " Origin of Species " Lamarck's hypothesis was not strenuously denied by naturalists. Early in the 'eighties, however, Weismann published his book on the germ-plasm, and the brilliancy and constructive ability of the speculations contained in this remarkable work, as well as the analogies which they suggested between organic and inorganic phenomena, compelled the attention of biologists. The essential parts of Weismann's hypothesis, as it was first presented to the world, are as follows : very early in the evolution of living from non-living matter many kinds of life-substance came into existence. These were chemical compounds of great complexity, able to accumulate and expend energy, and capable of indefinite growth and reproduction. They were able to exist in an environment which was hostile to them and which tended always to their dissolution, and which was able to modify their nature and their manner of reacting, though it could not destroy them. These elementary life-substances were very different from those which we know in the world of to-day. They were *naked* protoplasmic aggregates, undifferentiated into cellular or nuclear plasmata, much less into somatic and germinal tissues. All of their parts were similar, or rather their substance was homogeneous. But even with the evolution of the unicellular organism a profound change was initiated, for henceforth one part of the living entity, the nucleus, became charged with the function of reproduction, although it still continued to exercise general control over the functions of the extra-nuclear part of the cell. When the multi-cellular plant and animal became evolved, the heterogeneity of the parts of the organism became greater still. All the cells of the metazoan animal do indeed contain nuclei, but these structures are only the functional centres of the cells : some of the latter are sensory, others motor, others assimilatory, others excretory, and so on. Only in the nuclei which form the essential parts of the reproductive organs does the reproductive function persist in all its entire potentiality : there only does the protoplasm retain all the properties which were possessed by the primitive life-substance before it became heterogeneous, that is, before nucleus and cytoplasm evolved. When part of the primitive life-substance became secluded in a nuclear envelope, it became, to that extent, shielded from the action of the physical environment, and when the organism became composed of multicellular tissues this seclusion became more complete. Clothed in the garments of the flesh, it was henceforth protected from the shocks of the environment, and it became the immutable germ-plasm. But for a very long time before this evolution of tissues the naked life-substance had been exposed to the action of external physical agencies, and it had been modified by these into very numerous forms of protoplasmic matter. When multicellular plants and animals had been evolved there were, therefore, not one, but many kinds of life-substance in existence, and these have persisted until to-day as the unchanging germ-plasmata of the existing organisms.

The Weismannian hypothesis of to-day, supported and amplified, as it is, by subsidiary hypotheses, does not make the same appeal to the student as did the pristine and altogether attractive speculation of thirty years ago. The analogy which it then presented with the matured chemical theory of matter must have been almost irresistible. Just as the indefinitely numerous compounds of chemistry are only the permutations and combinations of some of eighty-odd different kinds of matter, so all the forms of life are combinations and permutations of some of the many different kinds of life-substance which came into existence before the evolution of the multicellular organism. And just as the chemical elements were regarded (in 1883) as immutable things, preserving their individuality even when they were associated together as compounds, so Weismann and his followers looked upon the different kinds of life-substance contained in the chromatic matter of the nucleus as immutable and immortal living entities. Associated together in indefinitely numerous ways by sexual conjugation, they may build up indefinitely variable living structures, but they remain individualised and lying side by side in the germ-plasmata of organisms, just as the atoms were supposed to lie side by side in the chemical molecule of the inorganic compound.

If these speculations were true, a change of morphology or functioning, acquired by the body, or somatoplasm, could not possibly be transmitted to the progeny of the organism, for by hypothesis the germ-plasm cannot be affected by external changes, and it is only the germ-plasm contained in the spermatozoon of the male parent, or in the ovum of the female, that shapes and builds the body of the offspring. As if this were not enough, Weismann and his followers argued that the transmissibility of a somatic change to the germ was inconceivable. Why ? Because the germ-cells are apparently simple : they are only semi-fluid protoplasmic cell bodies and nuclei, not differing appreciably from the cell bodies and nuclei of the somatoplasm (by hypothesis, it should be noted, the difference is profound). There are no structural connections — no nerves, for instance — which join together the cells of the bodily tissues with the parts of the germ and transmit changes in the former to the latter. How, then, could a somatic change affect the germ so that when the latter developed into an organism this particular change became repro-

---

[1] We know now that this statement is not quite accurate.

duced? Now this may have seemed·a conclusive argument in 1883, but is it so conclusive to-day? We know that the cells and tissues are not isolated particles, but that all are connected together by protoplasmic filaments. We know that specialised nervous tissues are not necessary for the transmission of an impulse from a sensory to a motor surface, but that such an impulse may be transmitted by undifferentiated protoplasm. We know that nerve-cells and nerve-fibres are not structurally continuous with each other but that the impulse leaps across gaps, so to speak. We know that events that occur in one part of the body of the mammal may affect other parts by means of the liberation of a chemical substance, or hormone, into the blood stream. It would be strange indeed if a logical hypothesis capable of accounting for the transmission of a particular change from the soma to the germ could not be elaborated.

But acquired characters were not really transmitted after all. So those who clung to Weismannism argued —an unnecessary task surely if this transmissibility were inconceivable. We cannot discuss the evidence here, and it is unnecessary that we should do so, since it is all considered in the popular books on heredity. There is an apparent consensus of opinion in these books which should not influence the reader unfamiliar with zoological literature, nor obscure the fact that many zoologists and botanists accept the opposite conclusion. The discussion is all very tiresome, but we may glean some results of positive value from it. It is unquestionable that very few conclusive and adequate investigations have been made : one cannot help noticing that the literature contains an amount of controversy out of all proportion to the amount of sound experimental and observational

P

work actually carried out. Most of the experiments deal with the consideration of traumatic lesions or mutilations, and it seems to be proved that such defects are not transmitted, or at least are very rarely transmitted. The tails of kittens have been cut off ; the ears of terrier-dogs have been lopped ; and the feet and waists of Chinese and European ladies have been compressed, and all throughout very numerous generations, yet these defects are not transmitted from parent to offspring. This kind of evidence forms the bulk of that which orthodox zoological opinion has adduced in favour of the belief in the non-inheritability of acquired characters, but does it all really matter ? What might be transmitted is a useful, purposeful modification of morphology, or functioning, or behaviour, induced by the environment throughout a number of generations—an adaptation rather than a harmful lesion. There is little conclusive evidence that such adaptations are inherited, though anyone who carefully studies the evidence in existence will not be likely to say that they are certainly not transmitted. Does, for instance, the blacksmith transmit his muscular shoulders and arms to his sons, or the pianiste her supple wrists and fingers to her daughters ? There are no observations and experiments in the literature worthy of the importance attaching to the question at issue.

It should be noted also that the germ-plasm is certainly not the immutable substance that the hypothesis originally postulated. Changes in the outer physical environment may certainly affect it ; thus the larvæ bred from animals which live in abnormal physical conditions (temperature, moisture, etc.) may differ morphologically from the larvæ bred from animals belonging to the same species but living in a normal environment. The latter must therefore react on the germ-plasm, but the environment formed by the bodily tissues which surround the germ-cells may also so react : thus the germ-cells may be affected by such bodily changes as differences in the supply of nutritive matter, for instance. The offspring may deviate from the parental structure as the result of structural modifications acquired by the parent during its own lifetime, and, even if the filial deviation were not of the same nature as the parental modification, its inheritance would be an adequate cause of *some* degree of transmutation.

It is, however, certainly difficult to prove that organisms transmit to their progeny the *same* kinds of deviation from the specific structure that they themselves acquire as the result of the action of the environment. Even if they did transmit such acquired deviations, it does not seem clear that this kind of inheritance alone would be a sufficient cause of the diversity of forms of life that we do actually observe in nature. Change of morphology would indeed occur, but we should expect to find insensible gradations of form and not individualised species. Let us suppose that Lamarckian inheritance acts for a considerable time on two or three originally distinct species inhabiting an isolated tract of land, and let us suppose that we investigate the variations occurring among all the organisms which are accessible to our observation with respect to some one variable character.

The diagram *A* represents what would seem to be the result of this process of transmutation. The numbers along the horizontal line are proportional to their distance from o, the origin, and represent the magnitude of the variation considered ; and the height of the vertical lines represents the number of organisms exhibiting each degree of variation. We should expect to find that all the variations were equally frequent in their occurrence, but this is not what a study of variability in such a case as we have supposed—that of the animals inhabiting an isolated part of land—does actually indicate. What we should find would be the conditions represented by the diagram *B*. There would be two or more *modes*, that is, values of the variable character which are represented by a greater number of individuals than any other value of the variation. The environmental conditions *favour* the individuals displaying this variation to a greater extent than they favour the rest.

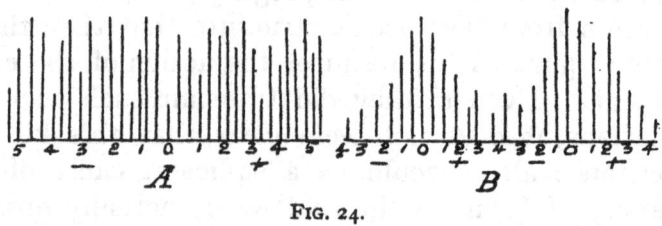

FIG. 24.

That is to say, the environment selects some kinds of variations among the many that are exhibited, and this is, of course, the essential feature of the hypothesis of the transmutation of species by means of natural selection of variable characters. Organisms enter the world differently endowed with the power of acting on the medium in which they live, or on the environment consisting of their fellow-organisms. Those that are most favourably endowed live longest and have a more numerous progeny than those that are less favourably endowed, and they transmit this favourable endowment to their offspring. Among the progeny of the progeny there may be some in which the favourable variation is still more favourable than it was when it first appeared. Thus the variations which are selected increase in amount. Elimination of the weakest occurs. The idea is eminently clear and simple, and possesses a great degree of generality : it is self-evident, says Driesch, meaning that it cannot be refuted, for it was certainly not clearly obvious to the naturalists before Darwin and Wallace. But, unless we choose to be dogmatic, we can hardly claim that it is an all-sufficient cause for the evolutionary process, and it is useless to attempt to minimise the difficulties of the hypothesis. It is not easy to make it account for the origin of instincts or tropisms, or for restitutions and regenerations of lost parts, or for the appearance of the first non-functional rudiments of organs which later become functional and useful. It is, indeed, possible to devise plausible hypotheses accounting for all these things in terms of natural selection, but each such subsidiary hypothesis loads the original one and weakens it to that extent.

Natural selection does not, of course, induce or evoke variations ; these are given to its activity, and they are the material on which it operates. What, then, is the nature of the deviations from the specific types of morphology that are selected or eliminated ? Not those induced by the environment, and transmitted in their nature and direction to the progeny of the organisms first displaying them. It is not unproved that such variations do occur, and it is even probable that they do occur. But we may conclude that the frequency of their occurrence is not great enough to afford sufficient material for natural selection. It is also clear that the ordinarily occurring variations that we observe in any large group of organisms collected at random are not alone the material for selection ; for we have seen that experimental breeding from such variations does not lead to the establishment of a stable race or "variety." Nevertheless some effect is produced, and this may be accounted for by supposing that the observed variations are really of two kinds—fluctuating variations, which are not inherited, and mutations, which are inherited. The small observed effect is due to the selection of the mutations alone : it is a real effect of selection, an undoubted transmutation of the specific form, but experimental and statistical investigations seem to show that selection from the variations that we usually observe is too slow a process to account for the existing forms of life.

Natural selection acts, therefore, on mutations. Now it seems that we are forced to recognise the existence of two categories of mutations, (1) those stable modifications of an "unit-character" which we term "Mendelian characters," and (2) those groups of stable modifications to which de Vries applied the term mutations. It seems at first difficult to see how permanent modifications of the specific form can be brought about by the transmission of Mendelian characters, for these characters are always transmitted in pairs. Let us take a concrete case—that of a man who has six fingers on his right hand, and let us suppose that this was a real, spontaneously appearing character or mutation which had not previously occurred in the ancestry of the man. Two contrasting characters would then be transmitted, (1) the normal five-fingered hand, and (2) the six-fingered hand. Both of these characters are supposed to be present at the same time in the organisation of the men and women of the family originating in this individual, but one of them is always latent or recessive. There would, however, be individuals in which only one of the characters would be present—either the normal or abnormal number of digits, but intermarriage with individuals belonging to the other pure strain would immediately lead again to the transmission of the contrasting characters, or allelomorphs, although marriage with an individual belonging to the same pure strain would carry on the normal or abnormal unmixed character into another generation. But if the possession of six fingers conveyed an undoubted advantage, and if natural selection did really act in civilised man as regards the transmission of morphological characters, then a stable variety (*Homo sapiens hexadactylus*, let us say) might be produced by its agency. The mutations which we consider in the investigation of the inheritance of alternating characters are therefore just as much the material for natural selections as the mutations which occur among the ordinary variations displayed by organisms in general : but since only one or two characters appear to be subject to this mode of transmission, the process would be so slow as to be inadmissible as an exclusive cause of evolution.

If we assume that de Vries' mutations are the material on which selection works, this difficulty is immediately removed, for we now have to deal with *groups* of stable deviations : not one or two, but *all* the characters of the organism appear to share in the mutability. But another difficulty now arises. A species of plant or animal may have got along very well with its ordinary structural endowment, and then a number of individuals begin to mutate. Some of the deviations from the specific type may be of real advantage, but others may not : we can, indeed, imagine an in-co-ordination between the mutating parts or organs which would be fatal to the animal ; on the other hand, there might be complete co-ordination, with the result that great advantage might be conferred upon the individual. It is easy to see how co-ordination of mutating parts is absolutely essential. An animal which preserves its existence by successful avoidance of its enemies would not be greatly benefited by a more transparent crystalline lens if the vitreous humour of its eye were slightly opaque ; and even if all the parts of the eye were perfectly co-ordinated, increased acuity of vision would not greatly help it if its limbs were not able to respond all the more quickly to the more acute sensation. Un-co-ordinated mutations would therefore tend to become eliminated, while co-ordinated ones would become selected and would become the characters of new species.

We must now ask why some groups of variations are co-ordinated while others are not, and it is here that we encounter the most formidable of the difficulties of any hypothesis of transformism which depends on the concept of natural selection. If we assume that the environment induces the appearance of variations, it seems to follow that these variations are likely to be co-ordinated, but we then invoke the principle of the acquirement of characters and their transmission by heredity. If, on the other hand, we assume that variations appear spontaneously, and quite irresponsibly, so to speak, in the germ-plasm of the organism, the selection, or elimination, by the environment will not occur until the co-ordinated or un-co-ordinated variations appear. It is far more likely that a large number of simultaneously appearing variations will be un-co-ordinated than that they will be co-ordinated. Merely as a matter of probability the progressive modification of a species will take place slowly—too slowly to account for what we see.

Two examples will make it easier to appreciate this difficulty. Evolution has undoubtedly proceeded in definite *directions*. There are two dominant groups of fishes, the Teleosts and the Elasmobranchs, and both must have originated from a common stock. All the characters in each kind of fish must have been useful (since they were selected), and all must have been modifications of the characters of the common stock. The latter became modified along two main lines, or directions, which are indicated by the characters of the existing Teleosts and Elasmobranchs. The whole skeleton, the gills, the circulatory system, and the brain differ in certain respects in these groups. Therefore a modification of the brain in the primitive Elasmobranchs was associated with a modification of the cranium, and therefore with the jaw-apparatus, and so with the branchial skeleton and the gills, and therefore also with the heart, and so on. Suppose that the evolutionary process included ten useful and co-ordinated variations—not an unlikely hypothesis—and suppose that each of these ten useful variations was associated with nineteen useless ones. The chance that any one of them did occur was therefore one in twenty ; and if they all occurred independently, that is, if the occurrence of any one of them was compatible with the occurrence of any other one, or of all the others, then the chance that all the ten variations occurred simultaneously was 20^{-10}, that is, one in the number 20 followed by 10 cyphers, a rather great improbability.

Most biological students are familiar with the similarity of the so-called eye of the mollusc Pecten and that of the vertebrate. The resemblance is one of general structure : in each of these organs there is a *camera obscura*, a transparent cornea, and behind that a crystalline lens. On the posterior wall of the camera there is a receptor organ, or retina, and this is composed of several layers of nervous elements. The actual nerve-endings are on the surface of the retina, which is turned away from the light, that is, the optic nerve runs towards the anterior surface of the retina, and then its fibres turn backwards. This " inversion of the retinal layers " occurs in all vertebrate animals, but it is exceptional in the invertebrates. The above general description applies equally well to the eye of the vertebrate and to that of Pecten.

Let us admit that these mantle organs in Pecten *are* eyes, for there is no conclusive experimental evidence that they really are visual organs, and plausible reasoning suggests that they may subserve other functions. Let us assume that the minute structure of the Pecten eye is similar to that of the vertebrate, and that its development is also similar : as a matter of fact both histology and embryology are different. Then we have to explain, on the principles of natural selection, the parallel evolution of similar structures along independent lines of descent ; for mollusc and vertebrate have certainly been evolved from some very remote common ancestor in which the eye could not have been more than a simple pigment spot with a special nerve termination behind it. In each case the organ was formed by a very great number of serially occurring variations, yet these two sets of variations must have been the same at each stage in two independently occurring processes. On any reasonable assumption as to the number of co-ordinated variations required, and their chances of occurrence, the mathematical improbability that these two series of variations did occur is so great as to amount to impossibility so far as our theory of transformism is concerned. Natural selection could not, therefore, have produced these two organs.

This argument of Bergson's fails, of course, in the

particular instance chosen by him, but this is because the case is an unfortunate one. Probably a morphologist could find a very much better case of convergent evolution—the parallelism between the teeth of some Marsupials and some Rodents, for instance. If detailed histological and embryological investigation should show a similarity of structure and development, in such compared organs Bergson's argument would retain all its force. We should then have to assume that there was a directing agency, or tendency in the organism, co-ordinating, or perhaps actually producing, variations.

Mechanistic biology can suggest no means whereby simultaneously occurring variations are co-ordinated : let us therefore think of these variations as occurring independently of each other, and let us ignore the difficulty of the infrequency of occurrence of suitably co-ordinated variations. Variations *are* exhibited by the evolving organism, and the selection of co-ordinated series is the work of the environment. But the environment is merely a passive agency, and it has to confer direction on the innumerable variations presented to it by the organism, rejecting most but selecting some. Let us think of the environment, says a critic of Bergson, as a blank wall against which numerous jets of sand are being projected. The jets scatter as they approach the wall : each of them represents the variations displayed by some organ or organ-system of an animal. Let us think of a pattern drawn on the wall in some kind of adhesive substance : where the wall is blank the sand would strike, but would fall off again, but it would adhere to the parts covered by the adhesive paint. The sand grains strike the wall from all sides, that is, their directions are un-co-ordinated. The wall is passive, yet a pattern is imprinted upon it.

From passivity and un-co-ordination come symmetry and order.

This argument withstands superficial examination, but to accept it is truly to be " fooled by a metaphor." *For what is the pattern on the wall?* It is the environment, says the critic. But what is the environment ? Inevitably we think of it as something that makes or moulds the organism, a way of regarding it that drags after it all the confusion of thought implied in the above analogy. Clearly the environment is made by the organism. Its *form,* that is, space, is only the mode of motion possible to the organism ; it is clear that whether the space perceived by an organism is one-, two-, or three-dimensional, space depends upon its mode of motion. Its universe is whatever it can act upon, actually or in contemplation. Atoms and molecules, planets and suns are its environment because it can in some measure act upon these bodies, or at least they can be made useful to it. Chloroform or saccharine, or methyl-blue and all the dye-stuffs prepared from coal-tar by the chemists, are part of our environment because we have *made them.* They existed only in potentiality prior to the development of organic chemistry. They were possible, but man had to assemble their elements before they became actual. In *making* them, he conferred *direction* on inorganic reactions.

Surely the organism itself selects the variations of structure and functioning that are exhibited by itself. If we hesitate to say that these modifications are creations, let us say that they are permutations of elements of structure, and that they were potential in the organisation of the creature exhibiting them. They *occur* in the latter if we must not say that they are produced. If they are detrimental, the organism is the less able to live and reproduce, and if it does reproduce, its progeny are subject to the same disability. If, as is usual, they simply do not matter, they may or may not affect the direction of evolution. If they are of advantage, that is, if they confer increased mastery over the environment, over the inert things with which the organism comes into contact, the latter enlarges its universe or environment, lives longer, and transmits to its progeny its increased powers of action. Indefinite increase of power over inert matter is potential in living things, and variation converts this potentiality into actuality.

This discussion is all very formal, but two conclusions emerge from it : (1) the insufficiency of the mechanistic hypotheses of transformism to account for all the diversity of life that has appeared on the earth during the limited period of time which physics allows for the evolutionary process. There does not appear to be any possibility of meeting this objection if we continue to adhere to the hypothesis of transformism already discussed : it faces us at every turn in our discussion. How great a part is played, for instance, by " pure chance " in the elimination of individual organisms during the struggle for existence ! Let us think of a shoal of sprats on which sea-birds are feeding : it is chance which determines whether the birds prey on one part of the shoal rather than another. Or let us think of the millions of young fishes that are left stranded on the sea-shore by the receding tide : it is chance that determines whether an individual fish will be left stranded in a shallow sandpool which dries up under the sun's rays, rather than in a deeper one that retains its water until the tide next flows over it. It is no use to urge that there is no such thing as " pure chance," and that what we so speak of is only the summation of a multitude of small independent causes. Let us grant this, and it still follows that the alternative of life or death to multitudes of organisms depends not upon their adaptability but upon minute un-co-ordinated causes which have nothing to do with their morphology or behaviour. These are instances among many others which will occur to the field naturalist : they shorten still further the time available for natural selection in the shaping of species, for they reduce the material on which this factor operates.

The other result of our discussion is to indicate that the problem of transformism of species is in reality the problem of organic variability. Let us assume that all the hypotheses of evolution are true : that the environment may induce changes of morphology and functioning in animals and plants, and that these changes themselves — the actual acquirements themselves, that is — are transmissible by heredity. Let us assume that the germ-cells may be affected by the environment, either the outer physical environment, or the inner somatic environment, and that mutations may thus arise. Let us assume that mutations may be selected in some way, so that specific discontinuities of structure—" individualised " categories of organisms, or species—may thus come into existence. Even then transformism is still as great a problem as ever, for the question of the mode of origin of these variations or modifications still presses for solution.

The simplest possible cases that we can think of present the most formidable difficulties. The muscles of the shoulders and arms of the blacksmith become bigger and stronger as the result of his activity. Why ? We say that the increased katabolism of the tissues causes a greater output of carbonic acid and other excretory substances, and that these stimulate certain cerebral centres, which in turn accelerate the rate of action of the heart and respiratory organs. An increased flow of nutritive matter and oxygen then traverses the blood-vessels in the muscles of the shoulders and arms, and the latter *grow*. Probably processes of this kind do occur, but to say that they do is not to give any real explanation of the hypertrophy of the musculature of the man's body, for what essentially occurs is the division of the nuclei and the formation of new muscle fibres. How precisely does an increased supply of nutritive matter cause these nuclei to divide and grow ? This is a relatively simple example of the adaptability of a single tissue-system to a change in the general bodily activity, that is to say it is a variation of structure induced by an environmental change.

In most cases, however, the variations of structure that form the starting-points of transmutation processes cannot clearly be related to environmental changes. Some fishes produce very great numbers of ova in single broods—a female ling, for instance, is said to spawn annually some eighteen millions of eggs. If we examine these ova we shall find that there is considerable variation in the diameter and in other measureable characters. We may attempt to correlate these deviations from the mean characters with environmental differences. All the eggs " mature," that is, they absorb water and swell, while various parts, such as the yolk, undergo chemical changes, during the month or so before the fish spawns. This process of maturation takes place in the closed ovarian sac ; and the eggs lie practically free in this sac, and are bathed in a fluid which exudes from the blood-vessels in its walls. It may indeed be the case that there are variations in the composition of this fluid in the different parts of the sac ; but these variations cannot be great ; the fluid is not really a nutritive one ; and the process of maturation is not hurried. We can hardly believe that the differences in morphology are due to these minute environmental differences. We may indeed say that we do not really study the germ cells when we measure the diameter of the egg or investigate any other measurable character, for the real germ-plasm is the chromatic matter of the nucleus. But this obviously begs the whole question : all the parts of the egg that are accessible to observation do vary, and ought we to conclude that the parts which are not accessible do not vary ? They *must* vary : the germ-plasm of each egg *must* be different from that of all the others, for the organisms which develop from these germs show inheritable differences. Further, can we contend that such minute environmental differences as we have indicated affect the germ-plasm ? Is it so susceptible to external changes ? A high degree of stability of the germ-plasm is postulated in the mechanistic hypothesis that we have considered, and indeed everything indicates that the specific organisation is very stable. Can it then be upset by such minute differences in the somatic environment ?

But the germ-plasm is not really simple, says Weismann ; it is a complex mixture of ancestral germ-plasms. The individual fish that we were considering arose from an aggregate of determinants, and half of these determinants were received from the male parent and half from the female one. But each of these parents also arose from a similar aggregate of determinants, which again were received from both parents, and so on throughout the ancestry of the fish. It is true that the germ-plasms contributed by the ancestors were not quite different, but they differed to some extent. Then there must have been as many permutations of determinants in the ovum from which the fish developed as there were permutations of characters in the eighteen millions of ova produced by it. Does not the hypothesis collapse by its own weight ?

It could only have been such difficulties as are here suggested that led Weismann to formulate his hythothesis of germinal selection. All those eighteen millions of eggs arose from the division of relatively few germ cells. Each of these original cells contained the specific assemblage of determinants, and the elements of the latter are of course the biophors. The biophors, it will be remembered, are either very complex chemical molecules, or aggregates of such. When the germ cells of the germinal epithelium divide to form those cells which are going to become the ova, the biophors must divide and grow to their former size, and again divide—it is really a chemical hypothesis that we are stating, though we have to employ language which seems to do violence to all sound chemical notions ! Now while the biophors were dividing and growing they were " competing " for the food matter which was in the liquid bathing them, and some got less, while others got more than the average quantity. In this way their characters became different, so that the eggs, on the attainment of maturity, became different from each other. Now, apart altogether from the impossibility of applying any test as to the objective reality of this hypothesis, it must be rejected, for it confers on bodies which belong to the order of molecules properties which are really those of aggregates of molecules. The typical properties of a gas, for instance, are not the properties of the molecules of which the gas is composed, but are statistical properties exhibited by

Q

aggregates of molecules.   On the hypothesis of germinal selection the properties of the animals which develop from the biophors are extended to the biophors themselves.   It was surely a desperate plight which evoked this notion !   It is, as William James said about Mr Bradley's intellectualism, mechanism *in extremis* !

We seem forced to the conclusion—and this is the result to which all this discussion is intended to approximate—that variations, heritable variations at least, arise spontaneously.   That is, there are organic differences which have no causes, a conclusion against which all our habits of reasoning rebel.   Yet it may be possible to argue that the problem of the causes of variations is really a pseudo-problem after all, and that there is no logical reason why we should be compelled to postulate such causes.   When we think of organic variability, do we not think, surreptitiously it may be, of something that varies, that is, something that ought to be immutable but which is compelled to deviate ?   But what is given to our observation is simply the variations among organisms.

Let us think of the crude minting machines of Tudor times which produced coins which were not very similar in weight and design.   From that time onward minting machines have continually been improved, each successive engine turning out coins more and more alike in every respect, so that we now possess machines which stamp out sovereigns as nearly as possible identical with each other.   Yet they are not quite alike, and this is because the action of the engine, in all its operations, is not invariably the same.   In imagination, however, we make a minting machine which does work perfectly, and turns out coins absolutely alike, but this ideal engine is only the conceptual limit to a series of machines each of which is more nearly perfect than was the last one. It is unlikely that matter possesses the rigidity and homogeneity which would enable us to obtain this perfect identity of result ; nevertheless this identity has a very obvious utility, and we strive after it, so that the result of our activity is the conception of a perfect mechanism, and of products which are identical. We assume that the reasons why our early and cruder machines were imperfect are also the reasons why our later and more perfect ones do not produce the results that we desire.

We are artisans first of all, and then philosophers, and so we extend this ingrained mechanism of the intellect into our speculations. To the biologist the organism is a mechanism which, in reproduction, *ought* to turn out perfect replicas of itself. It does not do so. Now, if biology shows us anything, it shows us that living matter is essentially " labile," that is, something fluent, while lifeless matter is essentially rigid, or nearly so. Yet, ignoring this difference, we expect from the organism that identity of result and operation that we conceptualise, but do not actually obtain from the artificial machine. We regard the organism, not only as a mechanism like the minting machine, but as the conceptual limit to a series of mechanisms. The reproductive apparatus of our fish does not turn out ova which are identical, but which differ from each other. Some of this variation, we say, is due to the action of the environment ; and some of it is due to the condition that each ovum receives a slightly different legacy of characters from the multitude of ancestors. The rest we conceive as due to the imperfect working of the reproductive machinery.

It is useful that science should so regard the working of the organism, for in the search for the causes of variation our analysis of the phenomena of life becomes more penetrating.   But does any result of investigation or reasoning justify us in assuming, as a matter of pure speculation, that deviations from the specific type of structure are physically determined in all their extent ? Have we not just as much justification for the belief that these deviations are truly spontaneous, and that they arise *de novo* ?   So we approach, from the point of view of experimental biology, Bergson's idea of Creative Evolution.

CHAPTER VII

THE MEANING OF EVOLUTION

APART from experimental investigation, the results of comparative anatomy, even if they are amplified by those of comparative embryology, and even if they include fossil as well as living organisms, do no more than suggest the occurrence of an evolutionary process. It is in vain that we attempt a demonstration of transmutation of forms of life by showing that a similarity of structure is to be observed in all animals belonging to the same group. We may show successfully that the skeleton of the limbs and limb-girdles of vertebrate animals is anatomically the same series of parts, whether it be the arms and legs of man, or the wings and legs of birds, or the pectoral and pelvic fins of fishes : such homologies as these were indeed suggested by the mediæval comparative anatomists apart altogether from any notions as to an evolutionary process. We may show that the simplicity of the skeleton of the head of man is apparent only, and that in it are to be traced most of the anatomical elements that enter into the skull and visceral arches of the fish ; and that fusions and losses and translocations of parts have occurred and can be made to account for the observed differences of form. All this might just as easily be explained by assuming a process of special creation, or the gradual development of a plan or design. Just as God made Eve from a superfluous rib taken from the body of her husband, so He may have formed the auditory ossicles of the higher vertebrate from those parts of the visceral arches of the lower forms which had become superfluous in the construction of the more highly organised creature. However much the language of evolution may force itself on biology, it does no more than symbolise the results of anatomy and embryology, and provide a convenient framework on which these may be arranged.

But if, as all modern experimental work shows, the form of the organism is, in the long run, the result of its interaction with the environment ; if, as indeed we see, this form is not an immutable one, but a stage in a flux ; and if deviations from it may occur with all the appearance of spontaneity, then it would appear that the observed facts of comparative anatomy and embryology are capable of only one explanation. They represent the results of an evolutionary process, and the relationships that morphological studies indicate are no longer merely logical, but really material ones. We can now endeavour to utilise these results in the attempt to trace the directions taken by the process of evolution.

In so doing we set up the schemes of phylogeny. We divide all organisms into plants and animals, and then we subdivide each of these kingdoms of life into a small number of sub-kingdoms, in each of which we set up classes, orders, families, genera, and species. But our classification is no longer merely a formal arrangement whereby we introduce order into the confusion of naturally occurring things. It is now a " family tree," and from it we attempt to deduce the descent of any one of the members represented in it.

The sub-kingdoms, or phyla, of organisms are the primary groups in this evolutionary classification. We divide all animals into about nine of these phyla—the Protozoa or unicellular organisms; the Porifera or sponges; the Cœlenterates, a group which includes all such organisms as Zoophytes, Corals, Sea-Anemones, and " Jelly-fishes; the Platyhelminth worms, that is the Tapeworms, Trematodes, and some other structurally similar animals which live freely in nature; the Annelids, a rather heterogeneous assemblage of creatures which includes all those animals commonly called worms; the Echinoderms, which are the Starfishes, Sea-Urchins, and Feather-Stars found in the sea; the Molluscs, that is the animals of which the Oyster, the Periwinkle, the Garden-Slug and the Octopus are good examples; the Arthropods, which include the Crustacea, the Insects, and the Spiders; and lastly the Vertebrates. Any such classification we naturally endeavour to make as complete a one as possible, but round the bases of the larger groups there cling small groups of organisms the precise relationships of which are doubtful. Yet, on the whole, these sub-kingdoms of organisms represent clearly the main directions along which the present complexity of animal structure has been evolved.

There is an essential structure which we endeavour to assign to all the animals of each phylum, and which is different from the structure of the animals belonging to all other phyla. The Protozoa, which for the present we regard as animals, are organisms the bodies of which consist of single cells. These cells may become aggregated into colonies, but they may as well exist apart from each other. They may be enclosed in limy, siliceous, or cellulose skeletons or shells, or they may possess limy or siliceous spicules in their tissues—these parts are non-essential, and the schematic Protozoan is a cell containing a single nucleus, and capable of independent existence. The Porifera, and all the other phyla, include organisms the bodies of which are made up of aggregates of cells. In the Porifera the cells, which are specially modified in structure, are arranged to form the internal walls of a "sponge-work" the cavities of which open to the outside by series of pores through which water is circulated. The bodies of the Cœlenterates are typically sacs formed by a double wall of cells—endoderm and ectoderm. This sac opens to the exterior by a single opening, or mouth, surrounded by a circlet of tentacles, and its cavity is the only one contained in the body of the animal. The Platyhelminth worms are animals the bodies of which are also composed of ectodermal and endodermal tissues, between which is intercalated another mesodermal tissue. They have a single digestive sac or alimentary canal opening to the exterior by means of a mouth only ; and they all possess a complex, hermaphrodite, reproductive apparatus. In all the other phyla there are also three principal layers or kinds of tissue, but in addition to the cavity of the alimentary canal there is also a body cavity, or cœlom, which is contained in the mesodermal tissues. The Echinoderms are such cœlomate animals, but the alimentary canal now opens to the exterior by means of both mouth and anus ; there are separate systems of vessels through which water and blood circulate ; the blood-vascular system of vessels is closed to the exterior, the water-vascular system being open ; and the integument is armed by means of calcareous spines or plates. The Annelids are animals with cylindrically shaped bodies, segmented so as to form numerous joints. Each segment bears spines or hairs or appendages of some sort, and also contains a separate nerve-centre. The alimentary canal opens externally by a mouth and anus, and there is a spacious body cavity. The Molluscs are unsegmented animals. The dorsal part of their bodies contains the viscera, and is protected by a shell; while the ventral part is modified for the purpose of locomotion. A fold of integument hangs down all round the body and encloses a cavity in which the gills are contained. The Arthropods are segmented animals. The body is armed by a calcareous carapace or shell which forms the exo-skeleton. Each bodily segment bears a pair of jointed appendages, and also contains a separate nerve-centre. The whole series of ganglia are connected together by means of a nerve-cord, and the nervous system lies ventral to the alimentary canal. The Vertebrata are also segmented animals, but the segmentation is not apparent externally. The skeleton is an internal one, and is built up round an axial rod or notochord. The nervous system is situated dorsally to the alimentary canal. There are two pairs of limbs.

Thus we set up an essential or schematic structure characteristic of each phylum. These schemata have no real existence : they are morphological types from which the actual bodily structure of the animals in each phylum may be deduced. They represent the minimum of parts which must be present in order that an animal may be placed in the phylum to which we assume that it may belong. But these anatomical parts need not actually be present in the fully developed organism : thus there are Crustacea in which the body is not segmented, and in which neither calcareous exo-skeleton nor jointed appendages are present ; and there are Vertebrata in which the limbs may be absent. But in such cases we require evidence that the essential anatomical characters which are absent in the fully developed animal have appeared at some stage in its ontogeny, and this evidence is usually available. Or if embryological evidence cannot be obtained, we require proof that the animal can be traced backwards in time, by means of other characters, to some form in which the missing structures reappear. The schemata are thus the generalised or conceptual morphology of the phyla. They are not the morphology of an individual organism, but they include the morphology of the race.

They are, Bergson says, themes on which innumerable variations have been constructed. Structural elements may be suppressed, as when the notochord disappears in the development of the individual Tunicate, though it is present in the larva. Or elements may disappear and become replaced by other structures, as when the true molluscan gills are lost in the Nudibranchs and are replaced by the respiratory plumes. They may be reduced to vestiges, as in the case of the " pen " of the Squids, or the " cuttlebone " of the cuttlefish, remnants of the domed shell of the primitive mollusc ; or in the appendix vermiformis of the human being, a remnant of the voluminous cæcum of the herbivorous animal. Structures which were originally distinct may coalesce, as when the greater number of the primitively distinct segments of the thorax of the crustacean fuse to form the " body " of the crab ; or when the segmental ganglia of the same animal fuse together to form the great thoracic nerve-centre. The form and situation of a structure may vary within wide limits : thus the digestive cavity of some Cœlenterates may be a simple sac, as in the Hydra, but it may be partially subdivided by numerous mesenteries as in the zooid of the Corals ; or the simple tubular alimentary canal in some Platyhelminth worms may be bifurcated in others, triple-branched in others again, or even provided with numerous lateral branches, as in the more specialised species in the group. Organs originally simple may undergo progressive modification : thus the eye of a mollusc may be a simple integumentary cavity in the floor of which there are some simple nerve-endings, and some black pigment ; or this cavity may close up so as to form a sac, and the anterior part of the sac may become transparent so as to form a cornea. Behind the cornea a lens may be formed, and the simple terminal twigs of the nerve-endings may become a many-layered retina of great complexity of structure. In the lowest Chordates the central part of the blood-vascular system is a simple contractile vessel, but this becomes the two-chambered heart of the fish, the three-chambered heart of the reptile, or the powerful four-chambered heart of the warm-blooded animal. Anatomical elements may change in function ; thus parts of the visceral skeleton in the fish may become the ossicles of the middle ear in the Reptiles and Mammals ; while its swim-bladder may possibly be represented in the higher vertebrates by the lungs.

Thus there may be suppression of parts leading to entire disappearance or to mere vestiges of the original morphology. A structure degenerating through disuse may become removed from its typical relations with other structures and may acquire altogether new ones. Or its increasing importance may lead to its hypertrophy and to increased complexity of structure, and perhaps to the inclusion of new anatomical elements, or to the incorporation of other parts, the function of which may originally have been quite different. In all sorts of ways organs and organ-systems may become anatomically different as the result of adaptive modifications, or indirectly as non-adaptive modifications induced by the adaptive modifications of adjacent parts. It is the task of comparative anatomy to trace these changes of morphology, aided by the study of embryology and by the comparison of the structure of the parts of fossil animals. Regarding the process of transformism as proved by experiments and observations in breeding and heredity, the naturalist endeavours to trace the lines along which evolution has proceeded from the results of morphological investigations.

Such results cannot have more than a very limited value, and it is often the case that several interpretations of morphological results are equally probable. We may conclude that the existing Teleost and Elasmobranch fishes are descended from a common stock which no longer exists; we may similarly conclude that the Birds and Reptiles are closely allied, more so than either group is to the Mammals; and we may conclude that the Primates—the group of Mammals to which Man belongs—is descended from some group allied to the existing Ungulates or Insectivores, while the Mammals themselves may have come down from some group of vertebrates related to both the Amphibia and the Reptiles. But as to the nature of the animals which combined the characters of the Birds and Reptiles, or of the Reptiles and Amphibia, we know nothing. Palæontology, if its results were more numerous than they are, would afford us the material for the discovery of these " missing links," and there can be no doubt that as the world becomes better known our knowledge of palæontological stages in the history of existing groups will become more complete, so that we may, in time, possess an actual historical record of the phylogeny of the main groups of animals. But it is remarkable that while the results of comparative anatomy and embryology, aided by those of palæontology, enable us to trace back short series of stages in the evolutionary process, they still show us gaps at all the places where lines of descent ought to converge. They show us, for instance, that the oldest Birds known were decidedly reptilian in their morphology, but they do not show us an animal which was neither Bird nor Reptile, but from which both groups of Vertebrata have descended ; and this is almost always the case in our hypothetical schemes of phylogeny. Morphology has continually to postulate the existence of "annectant" forms, "Archi-Mollusc," "Protosaurian," "Protochordate," etc. : hypothetical animals which combine the characters of those which lie near the bases of diverging lines of descent. There is nothing to guide us in the construction of these annectant forms except the progressive simplicity of structure indicated in the morphological and palæontological series. The earlier Birds had teeth, for instance, and so have the Reptiles, therefore the annectant form had teeth, and it was an animal combining the schematic morphology of both Birds and Reptiles. But just according to the value which we attach to one morphological character rather than another, so will the structure of the annectant form differ. Is, for instance, the alimentary canal of the Vertebrate the most fundamental and conservative part of its morphology : that is, is it the structure which has been most resistant to change in the course of the evolutionary process ? Then we may regard the Vertebrates as having descended from some animal which was closely related to the Annelid worms.

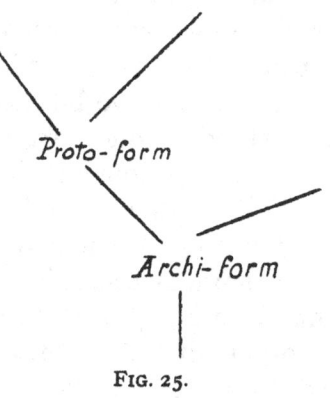

Fig. 25.

Or is the nervous system the most conservative part of the Vertebrate anatomy? If so, we may trace back the main Chordate stem to animals which included among their characters those of the most primitive Arthropods. In the one case the annectant form joins together the Vertebrate and Annelid stems, but in the other case it would join together the Vertebrate and Arthropod stems, a conclusion which a rigid application of the results of morphology would seem to make the more probable one.

But, however this may be, we must not fail to notice that annectant forms — " Archi-Mollusc " " Protosaurian," " Protochordate," and the like, are only fictions which we base on the precise importance that we attach to one part of the essential morphology of a group of animals rather than another. These hypothetical animals, and the genealogical schemes or phylogenies of which they form the roots, are conventional summaries of the results of comparative anatomy, this term being used to include the anatomy of the developing animal and that of extinct forms. So long as we do not possess a representative series of the fossil remains of the animals which have existed in the past, all schemes of descent founded on the comparison of the parts or the organs of living animals, or on the comparison of stages of development, must possess doubtful value when they profess to indicate the direction taken by evolution. Their true value lies rather in the way they epitomise our knowledge of morphology, and in the incentive which they give to sustained and minute investigation of the structure of animals.

Why did Haeckel's " Gastrea-Theorie " gain the acceptance that it did during the latter part of the nineteenth century? It correlated a great number of facts, in that it postulated a general uniformity of structure in the early developmental stages of very many animals belonging to widely separated groups. In all of these the ovum segments into a mass of cells, which then become arranged as a hollow ball (*A*). One side of this ball becomes pushed in so that the inner part of the hollow sphere becomes opposed to the inner wall of the upper part. Thus a little sac, consisting of two layers of cells, ectoderm and endoderm, and opening to the outside by an aperture, the blastopore, is formed (*B*). This is essentially the anatomy of the schematic Cœlenterate animal—*Hydra*, for instance, strongly suggests it. Suppose now that the lips of the blastopore fuse together at one place so that there are two openings into the cavity of the gastrula instead of one ; and suppose that the spherical organism elongates so as to form a cylinder, the elongation involving the fused part of the blastoporic region. Then we obviously have a worm-like animal with an alimentary canal, a mouth and an anus (*C*). Suppose further that an additional layer of cells becomes formed between the endoderm and ectoderm by proliferation from one of these tissues, and suppose that this becomes double and that a cavity appears between the two sheets of cells forming this middle layer : this cavity becomes the body cavity or cœlom (*D*). Now such blastula and gastrula stages appear in the ontogeny of animals belonging to widely different groups, and

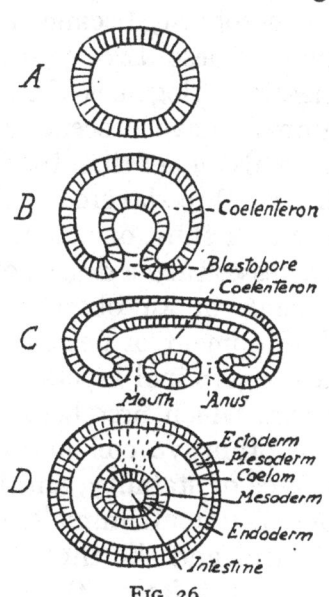

FIG. 26.

such a formation of the middle layer, or mesoblast, and of the mesoblastic or cœlomic cavities also actually occurs. Let us assume therefore that all multicellular animals have descended from a primitive Gastrea-form essentially similar in morphology to the gastrula larva ; and let us assume that all cœlomate animals have descended from a form in which a third layer of cells, or mesoblast, became intercalated between the other two. These two assumptions are the bases of the classic phylogenies of the last century ; all Cœlenterate animals have descended from a Gastrea-form, and all animals higher than the Cœlenterates have been evolved from a three-layered form. Implied in this hypothesis is also a third one, that the Gastrea-stage of evolution possesses such a degree of stability that it has persisted, though in an obscure condition it may be, in the development of nearly all multicellular animals. The triple germinal layers, endoderm, ectoderm, and meso-derm, which first became distinct from each other in the primitive cœlomate animal, also acquired a high degree of stability, and they have been transmitted by heredity to all animals higher than Cœlenterates. The Gastrea and the three germinal layers are therefore to be sought for in the developmental stages of all the higher animals, and they have usually been found. Let it be admitted that they may make a transient appearance—that they may be obscured in many ways, still they ought to be there.

The Gastrea-Theorie ceased to be useful, as a means of description, or a working hypothesis of investigation, after the rise of experimental embryology. It could not be proved that the process of development by gastrulation and the cleavage of a mesodermal layer are so very conservative that they have persisted throughout the greater part of the evolution of the animal world, yet without this proof it could not be contended that the veiled gastrula of the developing frog's egg, for instance, is related genetically to the gastrula of the Echinoderm larva. What experimental embryology does indicate is that the formation of gastrula and (in most groups) the three germinal layers are only the *means* of morphogenesis. In the division of the ovum, and the arrangement of the cells to form the organ-rudiments, the formation of the gastrula and the mesoderm are in general the line of least resistance in the process of development. If they do not appear, or are difficult to recognise in the ontogeny of a group of animals, it is not a sound method to assume their presence in an abbreviated or distorted form, postulating that they *ought* to be present, having been transmitted by heredity. Physical conditions undoubtedly influence developmental processes and there is no reason for assuming that all ontogenetic processes were originally the same.

If we do not strain the facts of our descriptions of organic nature, and if we do not build on unprovable conjectures, all that morphology certainly shows us is that the evolutionary process has led to the establishment of some dozen or so great groups of organisms, each with appended smaller groups more or less closely related to them. Whether these greater lines of descent are to be represented, as they usually are, as branches springing from a single stem, or whether they are truly collateral, each evolved independently of all the others, is a question which is not to be solved merely by the methods of comparative anatomy or embryology. The widely different, and equally probable, phylogenies of the past indicate that data for the solution of such a problem do not exist, not just yet at all events. What we may discuss with greater advantage is the question

R

as to which of the great subdivisions of life represents the main results of the evolution of complex organic entities from the simple living substances in which we suppose life first became materialised on our earth. What activities and structural forms represent the main manifestations of the evolutionary process ?

That is to say, what great groups of organisms are the *dominant* ones on the earth ? Greater or less degrees of dominance are indicated by the extent to which a group of organisms is distributed on the earth, by its abundance, and by the period of time during which it can be recognised in the fossil condition. Ubiquitous distribution implies a high degree of adaptability : a group of organisms inhabiting land and sea and atmosphere is obviously one in which the morphological structure has been elastic enough to admit of the development of various modes of locomotion ; and the limbs may be either the appendages of a terrestrial animal, or the fins, or other swimming organs, of an aquatic creature, or the wings of one adapted for flight. Dominance in this respect implies mobility and activity, and a relatively highly developed nervous system ; it implies the development of organs specialised for prehension, that is, for the capture of food ; and it also implies a high degree of adaptability to widely different physical conditions, to temperature changes, for instance. Dominance in geological time means also this great adaptability to changes in climatic conditions, and the development of means of distribution sufficient to overcome extensive physical changes on the surface of the earth. A terrestrial species might become isolated by the formation of a mountain range, or the submergence of the land adjacent to that which it inhabited, and some widely distributed species of plants and insects must have been able to traverse oceanic areas. The abundance of a group obviously implies great powers of reproduction,. the ability to withstand physical changes, and the ability to resist competition with other predatory creatures. Dominance, in short, means that the organism possesses in high degree the inherent powers of reproduction ; and also those activities which enable it to respond by adaptations of morphology, functioning, and behaviour, to environmental changes. These environmental changes are those which must have been experienced during lengthy geological periods, and also those experienced by the organism in its attempt continually to enlarge its area of distribution.

If we make a broad survey of the animal world we shall find that dominance in these respects has been acquired by three great groups of organisms, (1) the Bacteria, (2) the chlorophyllian organisms- (3) the Arthropods, and (4) the Vertebrates. In each case the threefold condition of wide distribution over all the earth, both in fresh and marine water areas, on the land and in the atmosphere ; of existence throughout the greater part of geological time ; and of ability to withstand environmental change, are satisfied. The bacteria are known to have existed in the carboniferous period. At the present time their distribution on the earth is universal : no part of the land surface, and no water masses, either marine or lacustrine—no matter how unsuitable they may be for the life of more highly organised creatures — are untenanted by bacteria. They are able to withstand extremes of temperature, or of salinity, which would be fatal to the multicellular plant or animal. Parasitism is a mode of life which they exhibit in a more manifold degree than do any other organsims. The upper regions of the atmosphere are the only parts of the earth and its envelopes which they do not inhabit.

The chlorophyllian organisms include those unicellular plants and animals—the distinction becomes obscure with regard to these organisms—which are pigmented blue, green, brown, or red owing to the existence in the cells of chlorophyll, or of some substance allied to this compound, and they include, of course, the green plants. Like the Bacteria their distribution is world-wide, extending over land and sea and freshwater areas ; and it is restricted mainly by the distribution of sunlight, and by a lower limit of temperature. The Marine Algæ, the Diatoms, the Peridinians, and other chlorophyll-containing organisms appear to inhabit all parts of the world ocean, certainly within a depth of about twenty to fifty fathoms from the surface of the sea. Green plants inhabit the land everywhere except within polar areas, the tops of high mountains, and over areas desert by reason of lack of water, or by the presence of mineral substances.

These conditions—temperature, light, soil, etc.—do not appear to limit the distribution of the Arthropods and Vertebrates. We find both kinds of animals in the deepest oceanic abysses (deep-sea fishes and Crustacea), in polar land and sea regions (Man, some Insects, Crustacea, and Birds), as well as in desert areas and on the summits of the loftiest mountains. The Ants share the subsoil with the Bacteria. Birds and Insects conquer the atmosphere by their activity and not, like the Bacteria, merely by being blown about. Crustaceans such as the Copepoda have much the same distribution in the sea as the Insects have in the atmosphere, while Isopods and Amphipods are a parallel, so far as the sea bottom is concerned, to the Spiders, Millipedes, and Ants on the land. Fishes are distributed throughout all depths, and in almost all physical conditions in the sea. Some species of marine Mammalia and Birds are quite cosmopolitan except that they are restricted to the upper layers of the ocean. Land Mammals are subject to the same restrictions as are the green plants, being unable to survive in desert and polar areas. The only parts of the sea which are not inhabited by Arthropods and Vertebrates are those limited deep strata of water (as in the case of the deeper layers of the Black Sea) where there are accumulations of poisonous chemical substances in solution. But the Bacteria inhabit even these regions.

Green plants, Arthropods, and Vertebrates appear as fossils in almost every part of the stratified rocks. The Trilobites represent the end of a long evolutionary process, and the same is to be said of the first fishes found in Silurian rocks, so that these groups of animals must have existed in the geological periods represented by those remains of rocks which are older than the earliest fossiliferous ones. Plant remains are present in Silurian rocks, but there can be no doubt that Ferns and other chlorophyllian organisms must have been in existence long before this time. We can hardly suppose that the Bacteria found in the Carboniferous rocks first appeared at this time in the earth's history : like the other great groups of life they probably had a prolonged history prior to that date of the geological formations in which they are first to be recognised. Our dominant groups of organisms may therefore be traced back almost to the very beginnings of life on the earth.

Dominance, such as we have defined it, cannot be said to have been attained by any other of the sub-kingdoms of life. Cœlenterates and sponges appear to have existed throughout the whole period during which the remains of organisms are to be traced in the rocks, but they have always been exclusively aquatic animals and they are very sparsely distributed in fresh water regions. Echinoderms are also a very old group, but they were more abundant in the past than they are now, and they appear to have been an exclusively marine group of animals. Molluscs have existed since the beginnings of stratified deposits and they are both aquatic and terrestrial animals, but they belong predominantly to the sea. They have always been relatively sluggish and inactive animals, with the exceptions of the great Squids and Cuttlefishes, but fortunately for the other inhabitants of the sea these formidable creatures appear to possess restricted powers of reproduction, and they have never been very abundant. All the smaller groups of animals are restricted in their distribution : the flat-worms occur sparingly both on the land and in the sea, and they attain their highest development as parasites in the bodies of other animals. Annelid worms, Gephyrea, Nemertine worms, Polyzoa, Rotifers, etc., are all groups of animals occurring mainly in fresh and sea water and none of them is abundant. Related to most of the great phyla are smaller groups : the extinct Trilobites, Eurypterids, etc., in relation to the Arthropoda ; the group represented now by *Peripatus* in relation to the Arthropods and Annelids ; the Enteropneusta and some other creatures which appear to possess affinities with the Echinoderms and Chordates ; and the extinct Ostracoderms, which appear to have been related to either the Arthropods or Vertebrates, or to both. All these smaller groups of animals we must regard as representing sidepaths taken by the evolutionary process—paths which have either ended blindly, as in the case of those groups which have become extinct, or which we can still trace in the existing remnants of groups which were formerly more abundant than they are now.

Only among the existing Bacteria, chlorophyllian organisms, Arthropods, and Vertebrates has the vital impetus found its most complete manifestation, and we may even narrow down the main path that evolution has taken to certain groups in each of these phyla. Some of the Bacteria—those which are exclusively parasitic in the bodies of the warm-blooded animals—have adopted a most specialised mode of life, and may even be said to exist only with difficulty, since the healthy animal is able to destroy them. Only those Bacteria living in the open or upon the dead tissues of plants and animals have attained to real dominance. Some green plants, like the Ferns, are far less abundant now than they were in the past ; while the Fungi and some other saprophytic and parasitic plants have specialised in much the same way as have the parasitic worms, and are restricted in their distribution. Marine Algæ are confined to a relatively narrow selvedge of sea round the land margin. The great trees, the grasses, and the microscopic green plants such as the Diatoms and Peridinians, represent the truly dominant organisms in the vegetable kingdom. On the side of the Arthropods and Vertebrates there have been many unsuccessful lines of evolution : the Trilobites, for instance, in the former group ; and the armoured Ganoid fishes, the armed Reptiles, the volant Reptiles, and the giant Saurians and Mammals among the Vertebrates. Among the existing Arthropods and Vertebrates there are some smaller groups which persist, so to speak, only with difficulty. Such are the Spiders, Mites, and Scorpions among the Arthropods ; and the Tunicates, the Dipnoan fishes, the tailed Amphibians, many Reptiles, and the volant Mammals among the Chordates : such are, of course, only instances of the less successful lines of evolution in these phyla. The dominant Arthropods and Vertebrates are the Crustacea, the Hymenopterous Insects, the Teleost and Elasmobranch fishes, and the terrestrial Mammals. The earth belongs to Man, to the social and solitary Ants, Wasps and Bees, the marine Crustacea, the Teleost fishes, the Trees, Grasses, and unicellular Diatoms and Peridinians, and to the putrefactive and prototrophic Bacteria. These are the organisms in which life has attained its fullest manifestations, and has been most successful in its mastery over inert matter.

In what kinds of activity and morphology, then, has the vital impetus found most complete expression ? We see at once that in relation to energetic processes life has followed two divergent lines—animal and vegetable. There is no absolute distinction between the energy-transformations which proceed in the living plant and animal—we return to this point later on— but we may trace an unmistakable difference in tendency, that is, in the direction taken by evolution. This difference we have already considered in an earlier chapter, but we may illustrate it by considering a lifeless earth, and also one tenanted only by plants, or animals, or by both.

In a lifeless earth all energetic processes would tend continually toward a condition of stability. The crust of the earth, that is, the part known to us by direct observation, is made up of rocks and the remains of rocks ; materials consisting of compounds of oxygen, silicon, iron, aluminium, sodium, potassium, calcium, and so on. They are substances which would be stable but for the eroding action of water, the gases of the atmosphere, and volcanic activity. But as volcanic

activity tends always toward cessation, the oxygen of
the atmosphere would gradually disappear, first by
its combination with oxidisable substances, and second
by its combination with the nitrogen of the atmosphere
under the influence of electric discharges. Carbon
dioxide would either combine with materials in the
rocks, or would remain in the atmosphere along with
nitrogen and other inert gases in a stable condition.
Water, moved by the tides and winds, would gradually
plane down the surface of the land, unless along with
other gases it would gradually become dissipated into
outer space. We see, then, that the materials of the
earth tend to fall into stable combinations, and that
they approximate toward conditions in which potential
chemical energy becomes reduced to a minimum, the
whole energy possessed by matter being that of the
motions of the molecules, that is, kinetic energy unavailable for transformations of any kind. It would
be an earth devoid of phenomena.

Vegetable life alone would be possible only for a
time on an earth such as we know it at present. The
green plant depends for its existence on the presence in
the soil of mineral substances such as salts of nitric
acid and of ammonia, and on the presence of water and
carbon dioxide in the atmosphere. The chlorophyllian
apparatus is essentially a mechanism whereby these
substances become built up into carbohydrates, like
starch and sugar ; hydrocarbons, like resins and oils ;
and proteids. The energy necessary for these syntheses
is obtained from solar radiation through the agency of
the chlorophyll plastids. The green plant would
depend for its supply of nitrate or ammonia on the
combination of the nitrogen of the atmosphere with
oxygen, or on the exhalations from volcanoes, and
these are irreversible processes which tend continually

toward cessation. The plant requires also carbon dioxide and the amount of this substance in the atmosphere is very limited, while the only inorganic source from which it can be renewed seems to be volcanic activity : this substance also would tend to disappear. A time would therefore come when plant life on the earth would cease to be possible because of the disappearance of the materials on which it depends ; but while it did exist its result would be the accumulation of chemical compounds of high potential energy. The result of the metabolism of the plant is the formation of such compounds as cellulose from woody tissues and shed leaves, of other plant carbohydrates, of oils and resins, and of proteids. In the absence of bacteria such substances would persist unchanged : even in an earth tenanted by bacteria such products as oils, lignite, peat, coal, etc., have been able to accumulate throughout geological time. The tendency of plant life is therefore toward the accumulation of compounds of high potential energy, and this process also is irreversible.

Bacterial activity would, of itself, make continued plant life possible on the earth. The essential characters of these organisms are their ability to bring about the most varied energy-transformations. From our present point of view bacteria may be divided into paratrophic, metatrophic, and prototrophic forms. Paratrophic bacteria are those which live as parasites within the living tissues of plants and animals : this mode of life is obligatory, and these organisms are unable to live in the open. The result of their activity is the breaking down of protoplasmic substance. Metatrophic bacteria are those that produce putrefaction and fermentation of organic compounds. They may be parasitic in their mode of life, but most of them live in soil, in water, and in the cavities of the animal body—the mouth, alimentary canal, nose, and vagina. Proteids are decomposed by them into simple chemical compounds such as amido-acids, and then these substances, along with carbohydrates, are fermented so as ultimately to form water, carbonic acid, and salts of nitric acid. These bacteria obtain their energy from the conversion of chemical compounds of high potential energy into compounds of low potential energy.   Prototrophic bacteria are never parasites, nor do they live in the cavities of the bodies of animals : they always live in the open. They carry on still further the action of the putrefactive bacteria by converting ammonia into nitrous acid, and nitrous acid into nitric acid.   Others reverse this series of changes by reducing nitric acid to nitrous acid, nitrous acid to ammonia, and ammonia to free nitrogen. Others again oxidise sulphuretted hydrogen to sulphuric acid, others ferrous hydrate to ferric hydrate, while it has recently been shown that some bacteria are apparently able to oxidise the carbon of coal to carbonic acid.   Some are able to oxidise the free nitrogen of the atmosphere into nitrous and nitric acids.   How precisely the energy necessary for these transformations is obtained is not at all clearly understood, and it may be possible that some of the prototrophic bacteria obtain their energy by making use of the un-co-ordinated kinetic energy of the medium in which they live.   From our point of view the net result of the activity of the predominant species of bacteria which inhabit the earth is that they *reverse* the processes which are the manifestations of the metabolism of plants and animals. The result of the metabolism of plants is the accumulation of stores of high potential compounds such as carbohydrates, and the depletion of the terrestrial stores of carbon dioxide and other materials necessary for the continued existence of the plants themselves. The result of the metabolism of the bacteria is the break-down of this accumulation of such compounds as carbohydrates, and the replenishing of the stores of carbon dioxide and nitrogenous mineral substance on which the plant depends. If bacteria are present, the life process becomes a reversible one.

Plant life and bacterial life are thus complementary to each other, for, on the whole, the energetic processes of the green plant proceed in the opposite direction to those of the bacteria. An organic world consisting of green plants and bacteria would therefore be one capable of permanent existence. Now, so far, we need only consider these various kinds of organisms as living protoplasmic substances in which energy-transformations of different types proceed. The bacterium is simply a cell containing a nucleus, and the green plant need only be a nucleated cell containing a chlorophyll plastid : this is, indeed, all that it is in the case of a Diatom or a Peridinian. The morphology of the green plant is only accessory to the chlorophyllian apparatus. Neglecting the reproductive apparatus, the higher green plant consists essentially of the chlorophyllian cells in the parenchyma of the leaf, for roots and stomata are only organs for the absorption of water and mineral salts from the soil and carbon dioxide from the atmosphere ; while the tissues of the trunk, stems, and branches are, in the main, apparatus for the conduction of these raw materials through the body of the plant, and, of course, the nutritive substances into which they are elaborated. All the innumerable variations of form in the plant (apart from the structure of the flower or other reproductive organ) are adaptations which provide for the absorption and distribution of these substances ; or for the mechanical support of the plant body; or are non-adaptive variations, pure luxuries, so to speak.

More than this is represented by the structure of the animal body, but we must first of all consider the points of difference between plant and animal regarded merely as apparatus in which energy-transformations occur. In the green plant energy is accumulated in the form of high potential chemical compounds, but in the animal energy is expended. Inorganic mineral substances are built up by the plant into carbohydrate, proteid, and fat or oil, but in the animal body carbohydrate, proteid, and fat are dissociated into water, carbonic acid, and urea (or some other nitrogenous excretory substance); and the urea or other analogous substance is broken down by bacteria into nitrate, water, and carbon dioxide. The metabolic activities of the animal are said to be "analytic" or destructive, while those of the plant are said to be "synthetic" or constructive, but these contrasting terms hardly describe accurately the essential nature of the activities of the two kinds of organisms. What further constitutes "animality"? It is *purposeful mobility*, and the energy-transformations that occur are the means whereby this mobility is attained. The plant is essentially immobile, for such movements as the turning of leaves toward the light, the down-growth of roots, the up-growth of stems, the twining of tendrils round supporting objects, and the opening and closing of flowers are only the movements of parts of the plant organism. They are constant, directed responses to external stimuli—real tropisms—and the extension of this kind of response so as to describe in general the movements of animals is only an instance of the insufficient analysis of facts. The movements of the typical green plant are therefore movements of its parts, they are few in number, they belong to a few simple types, and they are evoked by simple external physical changes in the medium. The movements of the typical animal are movements of the organism as a whole ; they are infinitely varied in their nature ; they are evoked by individualised stimuli and they are continually being modified by the experience of the organism.

The bodily structure of the animal is the means whereby this purposeful mobility is attained and the energy-transformations directed ; and the greater and more varied the movements of the animal, the more complex is its structure. In respect of the manner in which the energy-transformations are effected, that is, in respect of the material means whereby energy falls from a state of high potential to a state of low potential, the morphology of the animal is similar to that of the plant, that is, the energy-transformations are the functions of nucleated cells. But in the plant the kinetic energy of solar radiation passes into the potential energy of chemical compounds which become stored in the body of the plant ; while in the animal the potential energy of ingested chemical compounds passes into the kinetic energy of the movements of the animal itself. How exactly it moves, how this kinetic energy is employed is determined by the sensori-motor system.

It is the existence of the sensori-motor system that makes the animal an animal. What, then, is the sensori-motor system ? It is the skeleton and muscles, that is, the organs of locomotion, aggression, prehension, and mastication ; the peripheral sensory and motor nerves ; and the central nervous system or brain. The skeleton of an animal, whether it be the carapace or exoskeleton of a crustacean, or the vertebral column, limb-girdles, and limb-bones of a vertebrate, is a rigid and fixed series of supports to which the muscles are attached. Organs of locomotion are, for instance, the appendages of a crustacean, the wings of a bird or insect, the tail and fins of a fish, or the limbs of a vertebrate. Organs of aggression are the mandibles of a spider or blood-sucking fly, the chelate claws of a crab or lobster, the jaws of a fish, or the claws and teeth of a terrestrial vertebrate. Organs of prehension and mastication are in the main also those of aggression. All these parts consist of modified skeletal structures, teeth, claws, etc., attached to muscles which originate in the rigid parts of the skeleton. When we speak of the movements of an animal we speak of the motions of such parts as we have mentioned ; other parts do indeed move — the heart pulsates, the lungs dilate and contract, and the blood and other fluids circulate through closed vessels ; but these are movements of the parts of the animal, and are comparable rather with those movements of the plant organism that we have considered. They are not to be regarded as examples of the mobility of the animal in the sense of the exercise of its sensori-motor system.

A central and peripheral nervous system is, of course, bound up with a motor system. Receptor organs, eyes, olfactory, auditory, tactile organs of sense, and so on, are the means whereby the animal is *affected by* changes in its environment—it need not be cognisant of, or become aware of, or perceive these impressions on its receptor organs. These stimuli are transmitted along the sensory, or afferent, nerves to the central nervous system : this is the way in. The effector nervous organs are the motor plates, that is, the nervous structures in the muscles in which the nerves terminate.

The motor nerves are the efferent paths, the way out from the central nervous system.

The central nervous system is essentially the organ for the integration of the activities of the whole body. It is the " seat of multitudinous synapses," a description which better than any other applies to the morphology of the brain of the vertebrate animal. We have already considered what is meant by the term " reflex action," it is the series of processes which occur when a " reflex arc " becomes functionally active. A reflex arc consists of (1) a receptor organ, say a tactile corpuscle in the skin ; (2) an afferent nerve fibre ; (3) a nerve cell in the brain or spinal cord ; (4) an efferent nerve fibre ; and (5) an effector nerve organ, say a motor plate in a muscle fibre. The series of processes involved in a reflex action consist of the stimulation of the receptor organ, the passage of the afferent impulse into the brain or cord, the passage of the impulse through a series of cells in the nerve centre forming a synapse, the transmission of the impulse through the efferent nerve fibre into the effector organ in the muscle and the stimulation of the latter to an act of contraction. This is a purely schematic description of the structures and processes forming a reflex action and arc : in reality the path both into and out from the central nervous system is interrupted again and again, and at each place of interruption there are alternative paths. The interruptions occur at the synapses. At a synapse the nervous impulse passes through an arborescence of fine nervous twigs, into which the fibre breaks up, into a similar arborescence, and these two arborescences are not in actual physical contact : the impulse leaps over a gap. At numerous places in both brain and cord there are alternative synapses and at these places the impulse may travel in more than one direction.

The brain and cord are a switch-board of unimaginable complexity, so that an efferent impulse entering it from, say, the eye, can be shunted on to one nerve path after another, so that it may affect any muscle in the whole body. This is no fiction : it may actually be the case. In normal respiration a centre in the hind-brain is stimulated to rhythmical activity by the presence of carbon dioxide in the blood, and from it efferent impulses originate which stimulate the muscles of the chest wall and diaphragm. But in the distress of asphyxia every muscle of the body may be stimulated to activity in the effort to accelerate the oxygenation of the blood, and these are not spasmodic movements of the muscles of limbs, etc., but purposeful contractions having for their object the increased intake of air into the lungs. The central nervous system is, therefore, a switch-board—so mechanistic physiology teaches, neglecting any idea of an *operator*. But the whole trend of modern investigation is to show that every increase of specialisation in the evolution of the higher animal adds to the complexity of this nervous apparatus by increasing the number of alternative paths that an impulse originating anywhere in the body may take before it issues from the brain or spinal cord. Yet with all this increase of complexity it is nevertheless the case that in the higher animal the various parts of the central and peripheral nervous system are more and more integrated, so that in the actions of the animal it becomes more and more the organism *as a whole* that acts.

All other organs in the animal body—excepting always the reproductive apparatus—are accessory to the sensori-motor system. The alimentary canal and its glands dissolve the food-stuffs ingested ; the metabolic organs, that is, the cells of the wall of the intestine,

s

the liver, etc., transform these ingested proteids, fats, and carbohydrates of the food into the proteids, fats, and carbohydrates of the animal itself ; the heart, blood, and lymph vessels carry this food material to the muscles and nervous organs ; the respiratory organs absorb oxygen which is distributed throughout the body in the blood stream ; the execretory organs, that is, the lungs, skin, and kidneys, remove noxious materials like carbonic acid and urea, or its precursors ; and purposeful changes of functioning of all these organs are brought about by changes in motor activity. Round the sensori-motor system all the rest of the structure of the animal body is built up.

What we see clearly in the evolution of the animal body is the progressive increase of activity of the sensori-motor system. The *animal becomes more and more mobile.* It is in this way that dominance has been attained and all the directions of structural evolution in the past that have not tended in this direction have been unsuccessful, irreversible, evolutionary processes. Great size has not succeeded in the animal kingdom, and so the gigantic reptiles and mammals of the secondary and tertiary periods have become extinct. Defence against enemies by the development of dermal armour has not succeeded, and so the Dinosaurs, and other armed animals of the Tertiary Age have also become extinct. The transformation of the fore limbs of the reptile into wings, or the legs of the mammal into flappers, did not succeed, because all the rest of the structure of these animals had become adapted to locomotion on dry land, and the change of structure had become too profound to be modified : so the Pterodactyls passed away, as the whales of our own period are also passing. Only in the lightly boned, feathered bird, with the possibility of the development of powerful pectoral muscles, did indefinite possibilities of flight reside ; and only in the fish, with the concomitant evolution of gills, the reduction of a minimum of the mass of the alimentary canal and its glands, and the conversion of most of the muscles of the body into organs actuating the tail fin, was the completeness of adaptation to aquatic life realised. Mobility, a bodily structure capable of indefinitely varied movements, and a nervous system by the aid of which any part of the body might become linked to any other part—these were the structural adaptations that have been successful alike in Arthropod and Vertebrate.

There were apparently two main types of structure by means of which this mobility and elasticity could be attained, the Arthropod type and the Vertebrate type. There seems little to choose between them if we had to select one of them in order to obtain a highly mobile organic mechanism. Arthropod and Vertebrate seem to be equally complex if we take account of difference in size and the additional bodily mechanism that great size must involve. Certainly the musculature of the Vertebrate is more complex than in the Arthropod. But greater weight must require larger and more powerful muscles if the same degree of mobility relative to the size of the animal is to be attained, and this more complex musculature must carry with it a more complex brain. It must also be concomitant with a more massive skeleton, for rigid supports for the muscles must be present in the mechanism. Why are there no great insects or crustaceans ? Mr Wells has suggested in one of his novels the formidability of a wasp two feet long ! Such a creature would indeed be more dreadful than any predatory bird that we know if its activity were also that of the wasps that we know, just as a Copepod as large as a shark would be a more formidable animal than the fish. It seems possible that the reason for the smaller size of the Vertebrate is to be found in the nature of the skeleton. Powerful muscles would require a very strong and thick carapace, and this would attain a mass in a very large insect or crustacean which would require too much energy for its rapid transport. A rigid exoskeleton like that of an Arthropod also means that growth must take place by a process of ecdysis, that is, the animal grows only during the periods when it casts its shell ; and the necessity of this process of ecdysis must be a formidable disadvantage in the case of a very large animal, if indeed it would be possible at all. Thus the Arthropod developing an exoskeleton must remain small, and this smallness, fortunately for the Vertebrate, has made it the less formidable animal. It was an accident of evolution that the Arthropods developed an exoskeleton instead of an endoskeleton.

Undoubtedly the internal skeleton of the Vertebrates, with its light, hollow, cancellated bones, was mechanically the best means for the attachment of muscles. It made possible a greater degree of freedom of movement of the parts of the body, greater variety and plasticity of action, and it removed, to some extent, the limit of size and the embarrassing discontinuity of growth by ecdysis, with all the dangers that this involves. Above all, it led to the increased complexity of the central nervous system, since this became bound up with the increasing variety of bodily movement.

In the evolution of the dominant groups of organisms we see, then, the development of several tendencies. First, that tendency which seems to offer the greatest contrast to the universal tendency displayed in inorganic processes, the dissipation of energy. The plant organism is essentially a system in which energy is accumulated in the potential form. Then, in the animal kingdom we see that the main tendency of evolution has been the development of systems in which energy becomes expended in infinitely varied movements. It may seem, on superficial examination, that in the animal mode of metabolism energy is dissipated as it is in inorganic processes ; and this is the conclusion that we should reach if we considered the actions, and the results of the actions, of the lower animals only, that is, animals lower than man. We return to this point later on, but in the meantime it is to be noted that the fundamental division of organisms is that founded upon their activities as energy-transformers, that is, into plants and animals. Within each of these kingdoms of organisms structural evolution has occurred : the unicellular green plant has evolved along very numerous lines, each of them characterised by a different type of morphological structure. The unicellular animal has also evolved in a similar way with the result that the present phyla have become established. Looking at these great groups of animals, we see that two of them have attained dominance by the development along different lines of a sensori-motor system. Here we see another fundamental difference between the plant and animal organism, but one which is a consequence of the difference that exists between the two kingdoms in respect of the energy-transformations carried out by them. The plant is characterised by immobility, the animal by mobility.

Immobility implies unconsciousness, mobility consciousness, and this physical difference is the third one which we can establish between the plant and the animal. Now few physiologists are likely to accept this distinction as one which has any real objective meaning. Consciousness is not a concept to be dealt with in any process of reasoning, it is not even something felt in the way in which we speak of the feelings of pain, or light, or hunger : these are all states of our consciousness. The difference in ourselves, says Ladd, when we are sunk in sound dreamless sleep, and when we are in full waking activity, *that* is consciousness. If we reason about organisms and their activities as we do about inorganic things we have no right to speak about consciousness, for outside our own Ego it has no existence. The acting animal is only a body, or a system of bodies, moving in nature, and all its activities are to be described by a system of generalised force and position co-ordinates with reference to some arbitrarily chosen point of space. " This animal machine," says a zoologist, writing about instinct, " which I call my wife, exhibits certain facial contortions and emits certain articulate sounds which correspond with those emitted by myself when I have a headache, but I have no right to say that she has a headache." This kind of argument does not appear to be capable of refutation except, perhaps, by the domestic conflicts which it would usually evoke if applied in such cases as that quoted. In a description of nature by the methods and symbolism of science we see only systems of molecules in motion, and in those systems which we describe as organisms the motions are only more complex than they are in inorganic systems. Such is the method of science, as irrefutable in the study of the organism as we know that it is false. Valid in pure speculation according to the methods of the intellect it would nevertheless be absurd in the everyday affairs of common civilised life ; and the scientific man who applies it in his writing would nevertheless hesitate to apply it in the affairs of his own household.

We must recognise that our knowledge that other beings like ourselves, as well as animals lower in organisation than ourselves, are consciously acting organisms is intuitive knowledge, attainable because of community of organisation : our intuitive knowledge of the behaviour and feelings of our own brothers and sisters is greater than our knowledge of other men and women ; and we can, by intuition, place ourselves within the consciousness of an intelligent dog to a greater extent than in the case of other animals. This knowledge of the consciousness of other animals is not scientific knowledge and it is unattainable and unprovable by reasoning or methods of scientific observation. It is a conviction in itself incapable of analysis or proof, but yet a conviction on which we confidently base most of our dealings with our fellow-creatures, and which is justified by our experience.

It is nevertheless a scientific hypothesis of much the same validity as many other scientific hypotheses. We cannot bring ourselves to doubt that other men and women are consciously acting organisms, however impossible it may be to adduce scientific reason for the faith that is in us. We cannot doubt that a compass needle which " responds " by turning one or other of its poles towards us according as we push forwards one or other of the poles of a magnet is an unconscious piece of metal, though we find it impossible to say why this belief possesses such conviction. From this to the movements of the typical green plant is only a step. The turning of a green leaf towards the source of light, or the downward movement of a root into the soil, are responses to external stimuli which exhibit most of the inevitability of response of the magnet. They are " tropisms " : the plant leaf is obliged to turn towards the light so that the latter strikes against its surface perpendicularly, and the root must grow downwards because gravity acts along vertical lines. But suppose that reflex actions are tropistic : suppose, for instance, that the moth is bound to fly into the candle flame because the light stimulates both sides of its body equally and this orientates it and guides it towards the direction from which the stimulus proceeds. Complex actions, in the higher animal, on this view are chains of reflexes, and the acting must be unconscious and inevitable, just as the turning of the magnet or green leaf are unconscious movements. Therefore the actions of our fellow-creatures are unconscious and automatic, a conclusion toward which the whole tendency of mechanistic physiology forces us. Yet we know that the conclusion cannot be true.

Between the obligatory reaction of the compass needle to the magnet, or the analogous heliotropism and geotropism of the plant organism, and the infinitely variable responses of the higher animal toward changes in its environment, consciousness must come into existence. It is absent in the inorganic system and the typical green plant ; it is dim in the sedentary sea-anemone or mollusc ; it becomes brighter in the freely moving Arthropod or fish ; and it is most intense in man. This, it must be admitted, is only a belief, but accepting it as such we may attempt to support it by showing a parallelism of stages of structural complexity and actions. The sensori-motor system is absent in the green plant ; it is simple in the extreme in the sea-anemone ; and it is rudimentary or vestigial in the sedentary mollusc. It becomes more complex in the Arthropod or fish, and it is developed to the greatest degree in ourselves. If we now examine our own mental states, with their corresponding conditions of bodily activity, we see as clearly as possible that our consciousness waxes and wanes with our activities. It is absent in normal sleep, when bodily activity in the real sense ceases almost absolutely, when the cerebral cortex becomes inactive, and when the only movements performed are those truly automatic ones of parts of the body which are analogous to the movements of the plant organism. Such movements are the rhythmic ones of the heart and lungs, the movements of the blood, and so on, in general the movements leading to constructive metabolism. Consciousness is most intense in difficult unfamiliar actions : the lad learning to row ; the child learning scales on the piano, or the fingering of the violin ; the engineer assembling together the parts of a new machine ; or the artist engaged on a picture. In each of these cases the worker is acutely conscious, in a deliberative manner, of his own bodily actions. But with the habitual exercise of these movements, and with the ease and facility with which they are performed, consciousness that they are being performed fades towards nothingness.

What does this mean but that degrees of consciousness are parallel to degrees of complexity of deliberated and purposeful bodily movements or actions ? Or degrees of consciousness are also parallel to the attempt of the organism to perform these actions. What is pain, the most acutely felt of all our mental states ? It is, Bergson says, the consciousness of the persistent and unsuccessful effort of the tissues to respond purposefully to a persistently renewed stimulus. But complex actions require for their performance systems of skeletal and muscular parts capable of moving in the most varied ways, and a system of afferent and efferent nerves with all their connections in the central nervous system : that is, a sensori-motor system. Therefore just as the sensori-motor system is more or less complex so, in general, is consciousness more or less acute.

Yet in the same organism consciousness is the more or less acute as the actions which it performs are more or less familiar. The pianist who plays scales as a matter of exercise carries out most complex movements of hands and wrists unconsciously and without effort, but to play an unfamiliar composition for the first time without error involves attention of the highest degree. A girl who counts the sheets of paper coming from a machine seizes a handful in one hand, and drops a separate sheet between every two fingers of the other hand, repeating this most difficult operation with great rapidity, and counting the handfuls of sheets accurately while thinking and talking deliberately about some other matter. At the beginning of her work these actions were clumsily performed and facility was only attained by sustained attention to the movements of the hands, yet with experience they become unconsciously performed. Complex movements of the body and limbs and digits, involving the co-ordinated activity of numerous muscles, nerves, and nerve centres, are performed at first only after a high degree of conscious effort, but with each repetition of the series of movements the animal ceases to be aware of them, or at least of their difficulty. In the higher animals there are, therefore, two categories of actions, (1) those unfamiliar actions which are *difficult*, and in the performance of which the animal becomes conscious of complex muscular activities ; and (2) those habitual actions which have become *easy* by dint of repetition, and the performance of which is unattended by conscious effort. Analysis of our own activities reveals these two categories of actions, and we have no doubt whatever that the higher animals have the same feelings of difficulty and effort in the one case, and of lack of conscious effort in the other.

The difference is one of those which separate instinctive from intelligent activities. Now we hesitate to attempt the discussion of this much-controverted question of the distinction between instinct and intelligence : after reading much that has been said as to the nature of this difference, one rises with the uncomfortable impression that the time is not yet ripe for its discussion, and that the problem is still one far more for the naturalist than for the psychologist. Reliable data are still urgently required. Yet it is a question which we cannot fail to consider. The typical plant differs from the typical animal in that a sensori-motor system has been evolved in the one but not in the other ; and among the animals in which this system is developed to a high degree the activities which involve its exercise differ in their form. Actions of a stereotyped pattern characterise the behaviour of the higher Invertebrate, while in the higher Vertebrate all that we see indicates that the behaviour is the result of deliberation, and that the actions performed are not stereotyped but differ infinitely in their patterns. Just as clearly as differences in morphology differentiate Arthropod from Vertebrate, so also do differences in the mode of activity of the sensori-motor system mark divergent lines of evolution culminating in the Hymenopterous Insect on the one hand and in Man on the other.

What is the essential difference between an action performed instinctively and one performed intelligently ? It is not that the animal is unaware of its activity in the first case and not in the second ; however much we tend to " explain " organic activity in terms of inorganic reactions, we do not really believe that the instinctively acting wasp is a pure automaton, while admitting that the schoolgirl is acutely conscious of her own multifarious activities. It is not that the instinctive action displays a " finish," or perfection of technique, that the deliberative action lacks : the comb built by the wasp is not more perfect in its way than is the doorway constructed by a skilled mason, or the " buttonholes " stitched by a seamstress. It is not that instinctive actions are so absolutely stereotyped, as is sometimes assumed, while intelligent actions grow more perfect in their result by repetition : the work of the insect or bird is often faulty and it is improved by practice. The most obvious difference is that the instinctive action is *effective* the very first time it is performed, while the intelligent action only becomes effective after it has been attempted several times, or very many times, according to its difficulty. The flight of the young swallow is effective inasmuch as it sustains the bird in the air, but it is also an exceedingly difficult series of muscular efforts which is at first clumsily performed and which becomes more perfect by repetition. But the flight of an aeroplane, even now after years of experiment, is not always effective, and exhibits at its best all the imperfections of the flight of the young swallow. Yet can we doubt that in time it will exhibit all the ease and certainty and finish of the flight of the bird ?

The typical intelligently performed action is the action of a *tool*, or of a part of the body which is used for some other purpose than that which is indicated by its immediate evolutionary history, or by its previous use. The typical instinctively performed action is always the action of a bodily organ, the structure and immediate evolutionary history of which indicates that it originated as an adaptation for the performance of these particular actions, or category of actions.

Here it seems to us that we find the distinction between the two kinds of bodily activity ; and the distinction is one which depends for its validity on our notions as to what a tool is. An implement made by man is a piece of inert matter fashioned in order that it may be used for a definite preconceived purpose. It has an existence as a definite specific object *apart from its use* ; and its exercise by the man who made it and its existence in nature are two different things. Its use must be *learned*, and the results obtained by its employment become more perfect with every repetition of its use. But the mandibles of an insect are implements purposefully adapted for some action or series of actions, just as the pincers of the blacksmith are so adapted. They are, however, implements which are part of the organisation of the animal using them— organised tools—and it does not seem as if we ought to think of them, and of their shape and nature, as something apart from their exercise. Must we think of an animal as having to *learn* how to use any part of its body ? If so, then the problem of instinct remains with us in all its historic obscurity. But if we think of the existence of a bodily tool as something inseparable from the functioning of the tool, the problem becomes less obscure, or at least it can be stated in terms of some other problems which we have already considered.

We do actually think of bodily parts or organs as material structures quite apart from the consideration of their functions : it is the distinction between morphology and physiology—an altogether artificial one. An animal, for the morphologist, is a complex of skeleton, muscles, nerves, glands, and so on ; and it does not matter whether it is contained in a jar of methylated spirit or is running about in a cage. For the physiologist it is " something happening " ; but is it not really both things, and are not the structure and the functioning only two convenient, but arbitrary, aspects from which we consider the organism ? We ought not to think of diaphragm and lungs apart from the movements of these organs, and we do not say that the first breath drawn by the newly-born mammal is an instinctive action, involving the use of inborn bodily tools—the diaphragm, lungs, etc. We ought not to think of the lips and mouth and pharynx of the young baby apart from the actions of suckling the mammæ of its mother, but usually we say that this action is an instinctive one. Where does the ordinary functioning of an organ end and its instinctive functioning begin ? Are the muscular actions of the lobster when it frees its body and appendages from the carapace during the act of ecdysis instinctive ones ? Most zoologists would say that they are not, any more than the movements of the maxillipedes in respiration are instinctive ones, yet they probably would not hesitate to say that the action of the " soft " lobster in creeping into a rock crevice is instinctive. Does a young child really " learn " to walk ? It is more likely that the actions of walking are potential in its limbs and that they become actual when all the connections of nerve tracts and centres in its brain and spinal cord become established. What is the difference between the acquirement of the ability to walk and to write ? The latter series of actions are unfamiliar combinations of nervous and muscular activities which are no part of the organisation of the young child ; while the former are simply the result of the complete functional development of certain nervous and muscular apparatus.

It seems difficult, then, to express clearly what is

the essential difference between instinctive and intelligent behaviour ; and it is doubtless the case that
reasoned experiments and observations are still too
few to enable us to make sound deductions. But it
certainly seems as if we ought to think of instinctive
actions as having evolved concomitantly with the
structure of the organs which effect them : they are
those *inheritable adaptations of behaviour* which are
bound up with—are indeed the same things as—inheritable adaptations of structure. In performing them the
instinctively acting animal is doubtless aware of its
own activity, but we must think of this awareness as
being of much the same nature as our consciousness
of the automatic activities of our own bodies—the
rhythmic activities of the heart and respiratory organs,
or the actions of our arms and legs in walking, for
instance. It is knowledge of the inborn ability of the
organisms to use an inborn bodily tool.

In the intelligent action we certainly see something
different from this. The organ or organ-system which
carries out such an action functions in a manner which
is different from that for which it was evolved : the
action is the conscious adaptation of the organ for
some form of activity new to it, and this acquirement
of activity seems to be non-inheritable—at least it is
non-inheritable in the sense in which we speak of
acquired characters being non-inherited. It is accompanied, while it is being acquired, by a consciousness
which is deliberative, and is different from that awareness of its own activity which accompanies the acting
of the instinctive animal—the knowledge that it is
acting in an effective manner. It does not seem as if
the animal in so acting is aware of the relation of the
bodily tool to the object on which it is acting. But
intelligence seems to imply more than this : it implies

the knowledge of the organism that some parts of its body bear certain relations to the parts of the environment on which they are acting, and that these relations are variable ones and may be the objects of conscious choice.

CHAPTER VIII

IT is convenient that we should express the results of biological investigation in schemes of classification, for only in this way can we reduce the apparent chaos of naturally occurring organic things to order, and state our knowledge in such a way that it can easily be communicated to others. But we must always remember that the classifications of systematic biology are conceptual arrangements, depending for their precise nature on the point of view taken by their authors. The clear-cut distinctions that apparently separate phylum from phylum, class from class, order from order, and so on, do not really exist. There are no such categories of organisms in nature as genera, families, and the higher groupings. All that we can say exist naturally are the species, since all the organisms composing each of these groups are related together by ties of blood-relationship, and all are isolated from the organisms composing other species by physiological dissimilarities which render the plants or animals of one species infertile with those of any other. Such would doubtless have been the opinion of most botanists and zoologists prior to the work of de Vries, but we must now recognise that the systematic, or Linnean, species of the nineteenth century was just as artificial a category as were the genera and families. Our arrangements of plants and animals into systematic

T

species and the higher groupings are therefore convenient ways of symbolising the results of morphological and physiological investigations, although they also indicate the main directions taken by the evolutionary process, but the manner in which they are stated in taxonomic schemes is always a more or less formal one.

There are no absolute distinctions between group and group, even between the animals and the plants. There is nothing, for instance, in the morphology of a Diatom to indicate that it belongs to the vegetable kingdom, or in that of a Radiolarian, to indicate that it is an animal. Peridinians are either plants or animals according to the general argument, or the point of view of the author who writes about them. Even a study of the energy-transformations that are effected in the living substance of these lower organisms does not afford an absolute distinction : synthetic metabolic processes in which energy passes into the potential condition may be carried out in animals, while many plants—the saprophytic fungi, or the insectivorous plants, for instances—may effect analytic energy-transformations of essentially the same nature as those exhibited in the typical mode of animal metabolism. Motility and the possession of a sensori-motor system do not afford the means of making a sharply drawn line of division between plants and animals. Potential energy passes into the condition of kinetic energy in the typical animal, and this kinetic energy is directed by the sensori-motor system. But some lower unicellular plants are motile, and they possess the rudiments of a sensori-motor system in the flagella by which their movements are effected. On the other hand, the sensori-motor system has become vestigial in many animal parasites—in the Crustacean *Sacculina*, for instance, which is parasitic on some Crabs. The possession of consciousness, in so far as we can say that other animals than ourselves possess it, is no distinction between the two kingdoms of life. Consciousness, judged by the degree of development of motility, must be supposed to be absent or very dim in the extreme cases of parasitism attained by some animals ; on the other hand, we may assume that it is present, to some extent at least, in the highly motile zoospores of the Algæ. Thus some lower organisms, the Peridinians and the algal spores, exhibit all the characters which we utilise in separating animals from plants—the chlorophyllian apparatus, by means of which the kinetic energy of solar radiation becomes transformed into the potential energy of organic chemical compounds ; the apparatus of receptor and motile organs, by means of which the potential energy of stored chemical compounds passes into the kinetic energy of bodily movements ; and the existence (so far as we can say that it exists in organisms other than ourselves) of some degree of consciousness.

Neither do those morphological schemata which we construct as diagnostic of phyla, or classes, or orders, etc., separate these groups from each other so clearly and unequivocally as our classifications suggest. It might seem for instance that the presence or absence of a notochord would sharply distinguish between the vertebrate and invertebrate, but structures which suggest in their development the true notochordal skeleton of the typical vertebrate animal are to be traced in animals which exhibit few or none of the characters which we regard as diagnostic of the Vertebrate. Typical Arthropods and typical Vertebrates seem to be distinct from each other, but the extinct Ostracoderms of Silurian times *may* have been animals which possessed an internal axial skeleton, and which were also armed by a heavy dermal exo-skeleton. It is a hypothesis of considerable plausibility that they really were Arthropods, on the other hand they are usually regarded as Vertebrates. So also with most other phyla : the morphological characters which absolutely distinguish between one group and others are very few indeed, and the small appended groups that lie about the bases of these larger groups may present one or other of the characters of several phyla. Looking at the morphology of the animal kingdom in a general kind of way, one does indeed see that a certain structural plan is characteristic of the organisms belonging to each of the great phyla, while more detailed structural plans may be said to be characteristic of the sub-groups. But minute morphological and embryological investigation reduces almost to nothing the characters which are absolutely diagnostic of these various groups.

No more than the nature of the energy-transformations, and the essential morphology, does the behaviour of animals afford us the means of setting up absolute distinctions between group and group. Really tropistic behaviour is exhibited by the movements of the stems, roots, and leaves of green plants, or in the movements of Bacteria, and perhaps some unicellular animals. Typically instinctive behaviour is exhibited by the individuals of societies of Insects and by many solitary-living animals belonging to this class ; and typically intelligent behaviour is exhibited by the acting of the higher Mammalia. Yet there is undoubtedly much that is truly instinctive in the behaviour of Man, and something of the same nature as his intelligence seems to inhere in the instinctively-acting mammal or insect : how else could an instinctive action become capable of improvement ? We cannot doubt that intelligence is manifested by a dog or by much that we see in the behaviour of ants. No rigid distinctions between tropisms, such as we have mentioned above, and the reflexes that may be taken to constitute instinctive behaviour, can be established. Minute analysis, such as that carried out by Jennings on the swimming movements of the Protozoa, leaves us quite in doubt as to how these modes of behaviour are most properly to be described; and all the controversy as to the nature of tropisms, reflexes, instinct, and intelligence surely indicates that these modes of behaviour have something that is common to all of them, and that no clear and certain distinction can be said to separate one from the other. Even those psychic processes which we call intellectual do not seem to be different in kind from some that we attribute to the lower animals: the Protozoan *Paramœcium* studied by Jennings, or the crabs, crayfishes, and starfishes studied by Yerkes and others really *learn* to perform actions, but this learning is said to be the result of a process of "trial and error." The animal tries one series of movements and finds that it fails, tries another and another with a similar result, and in the end finds one that is effective. This is remembered, and when the same problem again confronts the animal it is solved after fewer trials, and finally, after experience, the end-result is attained at once without previous trials. Now many of what we call truly intellectual processes are certainly processes of precisely this nature. Hypothesis after hypothesis *occurs* to the scientific man (or to the detective, or to the engineer confronted with some exceptional difficulty), and one after another is tested by actual trial, or by a process of reasoning (which is really the rapid and formal resuming of previous experience), until a hypothesis verifiable, or *a priori* verifiable, is found. What, for instance, are our mathematical methods of integrating a function, or working a long division sum, but methods of scientific " guessing," and verification of the hypotheses so made ? They are truly instances of the method of trial and error practised by the lower animals.

All the above amounts to saying that there is a community of energetic processes, of morphology, and of behaviour in animals and plants. " Protoplasm " is the same, or much the same chemical aggregate, whether it is contained in the cells of animals or plants. The cell, with its nucleus, chromatic architecture, cell-inclusions, and cell-wall, is essentially the same structure in all organisms. The complex and specialised process of nuclear division in tissue growth, or the series of events which constitute the acts of fertilisation of the ovum or its plant correlative, are the same all through the organic world. The sensori-motor system—receptor organ, nerve-fibre and cell, and effector-motor organ—is the same all through the animal kingdom. Alimentary canal and glands, enzymes, excretory tubules, contractile blood-vascular apparatus—all these are structures which are functionally the same, which are built on essentially the same morphological plan. Life, whether it is the life of plant or animal, makes use of the same material means of perpetuating itself on the earth and avoiding the descent of matter towards complete inertia.

Absolute dissimilarities, dissimilarities such as those between atoms of hydrogen and oxygen, or between a point and a straight line, or between rest and motion, do not exist between the different categories of entities that make up the organic world. Yet differences do exist, and must we conclude that because these differences are not absolute ones, because they are differences of degree, and not of kind, they are not essential, are not differences at all ? Must we say, for instance, that although an animal is a much more efficient machine than a gas-engine (in the sense of efficiency as understood by the engineer), there is really no difference between them, that they are both thermodynamic mechanisms, since in both energy is dissipated ? Ought we to say that, because the last steps in the formation of urea in the animal body are synthetic ones, there is really no difference between the nature of the energy-transformations that occur in the animal and the plant modes of metabolism ? Ought we to say that, because a dog may sometimes act intelligently and a man instinctively, psychically they are similarly-behaving organisms ? Surely this amounts to saying that, because things are not absolutely different, they are the same ; and surely the mode of reasoning is a vicious one!

What we clearly see in the different kinds of organisms—in the metabolically constructive plant and the metabolically destructive animal ; or in the instinctively-acting Arthropod and the intelligently-acting Mammal—is the progressive development of different *tendencies*. If the green plant is, in its essence, the same kind of physico-chemical constellation as is the animal, yet the tendency of its evolution has been that more and more it has acquired the habit, or the power, of using solar radiation to combine together carbon dioxide, water, and nitrogenous inorganic salts to form proteid and carbohydrate substances. On the other hand, the tendency of the animal has been more and more to absorb into its own tissues the proteid and carbohydrates synthesised by the green plant, and then to break these substances down into carbon dioxide and water, and less and less to effect such syntheses as are effected by the plant. Even if the Annelid worm, the Arthropod, and the Vertebrate were, at the origin of their ancestries, animals which were very like each other in the morphological sense ; even if there are some Arthropods which are very like Annelids, and some Annelids which might very easily be imagined to become transformed into Vertebrates, and some extinct Arthropods which may after all have been Vertebrates, yet it is the case that the tendencies of the evolution of each of these groups have been very different. All the while the Vertebrate tended more and more to develop a rigid axial rod or notochord, becoming later a jointed vertebral column, and a soft, pliable, exo-skeleton ; while the Arthropod tended more and more to develop a rigid exo-skeleton, and to remain soft in its axial parts. Even if these two tendencies may not have been fully realised, is it not the case that they are really different things ? The evolutionary process has therefore been, in its essence, the development, or unfolding, of tendencies originally one.

What is the evolutionary process ? It is usually regarded as a progress from organic simplicity towards organic complexity. Yet if we think about it as a physical process we cannot say that any one stage is any more simple or complex than any other stage. Let us compare organic evolution with the process of inorganic evolution, as, of course, we are compelled to do if we regard the former process as a physico-chemical one. Assume, then, that the nebular hypothesis of Kant and Laplace is true—it will make no difference to our argument even if this hypothesis is not true, and it is more easily understood than any other hypothesis of planetary evolution. Originally all the materials composing our solar system existed in the form of a gaseous nebula possessing a slow rotatory motion of its own.   It does not matter that the silicates, carbonates, oxides, and all other mineral substances that we now know existed then in the form of chemical elements, or the precursors of chemical elements : all the material bodies now present in the solar system were present in the original nebula.   The energy of this nebula consisted of the potential energy represented by the separation of atoms which later on became combined together, and of the kinetic energy of motion of these atoms ; and this material and energy, together with the other cosmic bodies radiating energy to it and those bodies receiving the energy which it lost by its own radiation, constituted a system, in the sense of the term as it is employed by the physicists.   Now, in the process of cosmic evolution this system became transformed, because it was continually losing energy by radiation.   As it cooled, the mean free paths of its atoms and molecules became less and less, and finally condensation to the liquid and then to the solid condition occurred.   The parts of the nebula continually gravitated together, so that it became smaller and smaller while its rotatory motion became greater. Finally, mechanical strains became set up in its mass as the consequence of the increased velocity of rotation, and disruption occurred with the formation of the sun, the planets, and the satellites.   There was no increase of complexity of the system.   At any moment of time its elements, that is, the chemical atoms composing it and the energy of these atoms, was the same as at any other moment of time.   Heat-energy may have been radiated from one part of the system—the heated nebula—to some other part of the system—the other cosmic bodies absorbing this radiation, but the total energy of the system remained the same.   The chemical atoms may have combined together to form molecules and compounds, and their energy of position may have become the energy of motion, but the ultimate materials were still the same. What happened during the cooling and contraction of the nebula was a rearrangement of the elements of the system, that is, of the atoms and their energies. At any moment of time the condition of the system was an inevitable consequence of the condition at the moment immediately preceding this, and a strict functionality, in the mathematical sense, existed between the two conditions. It was not more complex in the later stage than in the earlier one—it was merely different. Stages of evolution were really *phases* in a transforming system of matter and energies.

If we choose to regard organic evolution as a similar process of physico-chemical transformation, we must also regard the totality of life on our earth, with all the inorganic materials which interact with organic things, and with all the energies, cosmic and terrestrial, which also so interact, as a system in the physical sense. We are now compelled to think about this system in the same way as we thought about the cosmic one, that is, we must postulate that a rigid mathematical functionality existed between any two conditions of it, and that the latter condition was inevitably determined by the former one. We must think of the system as at all times composed of the same elements. In its later condition life may have been manifested in a greater mass of material substance than in its earlier conditions, but this increase of mass was only the increase of one part of the system at the expense of another part. At all times, then, the constitution of the system was the same, and different stages of the evolutionary process have only been different phases, or arrangements, of the same elements. At no time was the organic world any more or less complex than at any other time.   In its " primitive " condition *all was given.*

Mechanistic biology does not, of course, hesitate to accept this view of the evolutionary process. The " Laplacian mind " must have been able to calculate what would be the condition of the system at any phase, knowing the positions of all the atoms or molecules in the original nebula, and the velocities and directions of motions of all these atoms or molecules. Just as (in Huxley's illustration) a physicist is able to calculate what will be the fate of a man's breath on a frosty day, so the Laplacian mind must have been able to predict the fauna and flora of the world in the year 1913 from a complete knowledge of the material nature and energetic properties of the nebula from which it arose.

We cannot fail to see, on reflection, to what this view of the nature of the evolutionary process leads us. The primitive world-nebula was a system of parts which had extension in space. Materially it consisted of atoms isolated from each other by space, and energetically it consisted of the movements of these atoms, and of the energy of their positions with regard to each other. No two atoms could occupy the same space— they mutually excluded each other : this is what we mean by saying that the original—and every other— state of the system was a state of material things or elements spatially extended. Therefore, if the physical analogy is consistently to be retained, the organic system undergoing evolution was a system of elements which at any moment whatever were spatially extended. It was really a system of atoms or molecules possessing kinetic energy of motion, or potential energy of position —molecules which lay outside each other, and energies which were really the movements or positions of these molecules, and which therefore lay outside each other in the same sense.

The evolution of the individual organism must be a process of the same kind. Like cosmic and phylogenetic evolution, it is apparently a progress from the simple to the complex. A minute fragment of protoplasmic matter, homogeneous in composition, or apparently so, grows and differentiates, becoming the complex structure of the adult organism. Here the system in the physical sense is the fertilised ovum, the oxygen and nutritive matter which have become incorporated with it, and the physical environment with which these things interact. All these elements existed in that phase of the system which contained among its parts the fertilised ovum, as well as in that phase which contained the fully developed organism. Complex by comparison with the fertilised ovum and its environment as the adult animal and its environment may seem to be, it is only a different phase of the same system. Further, all the parts that form the tissues of the adult, and all their motions, are spatially extended, and are only rearrangements of the molecules and of the motions of the molecules that were actually present in the system in its initial phase. Speculation along these lines has led to all the results of Weismannism. All the parts of the adult organism are really present in the fertilised ovum and the nutritive matter which is to build up the fully developed animal, not in potentiality it must be noted, but actually present in the spatially extended condition. It is true that the hypothesis only requires that the determinants of the adult organs and tissues, and of the adult qualities, should be present in the ovum ; but since the energies necessary for the separation of these determinants, and for their arrangement and growth in mass, must also be

present in the initial phase of the system, it is evident
that the hypothesis implies that all the material
structure of the animal is present in the spatially extended form in the initial phase of the system.   Just
as the adult animal is a manifoldness of material parts
and energies that possess extension, so also is the undifferentiated embryo and its material environment an
extensive manifoldness.   We cannot otherwise conceive
it if we are to retain the mechanistic view of the development of the individual organism.

Let us think of the process of organic evolution in
another way by comparing it with the mathematical
process by which we form the permutations and combinations of a number of different things.   Individual
development is termed the assumption of a mosaic
structure, that is, all the parts of the adult are assumed
to be present in the embryo, but in a sort of " jumbled-
up " condition.   As development proceeds, these parts
become sorted out and arranged in a pattern which
continually becomes more and more distinct.   Much
the same process of arrangement and segregation must
be assumed to have occurred during the process of
racial evolution :   the parts of the " primitive " life-
substance, with all the parts of the physical environment which become incorporated with it during its
evolution, must have become segregated and arranged
so as to form the existing species of plants and animals.
A permutation, then, of the separate things a, b, c—
x, y, z, is an arrangement of all these things :   obviously
there are a very great number of ways in which the
letters of the alphabet may be arranged, $\lfloor 26$ in all.   But
we may take some of the letters and arrange them in
different ways :   the selections a, b, c, d, can be arranged
in $\lfloor 4$ ways; b, c, d, e, also in $\lfloor 4$ ways, and so on.   Thus
by a process of dissociation and arrangement of a certain

number of elements, a very great number of different things—things which consist of elements spatially extended—can be obtained.

The group of things, $a, b, c, d—x, y, z$, was an extensive manifoldness, since it was formed by juxtaposing in space the separate units of which it is composed. Yet it is an unitary thing, for it is a different thing from the group, $b, c, a—x, y, z$. It is also a multiplicity, for it can be transformed into every one of the |26 permutations, and broken up into the selections of some of the separate things of which it is composed, and of the permutations of the things taken in each of these selections. In a way these arrangements exist in the group $a, b, c—x, y, z$, and yet the group itself possesses no other actual extended existence than the group of things that it is. It is an *intensive multiplicity or manifoldness* in that the potentiality of all the arrangements exists in it but not in the spatially extended condition. It is a multiplicity only when we associate with it the mental operations by which we conceive of its dissociation and rearrangement. By reason of these mental operations the intensive multiplicity of the group becomes the extensive multiplicity of its arrangements.

This appears to be the only really philosophical way in which we can attempt to picture to ourselves the processes of individual and racial evolution. The " primitive " life-substance, or the undifferentiated ovum, each of them with its environment, was an intensive manifoldness, a multiplicity of distinct things or qualities which co-existed, and which were not separate each from other in that they occupied different compartments of space, but which interpenetrated each other. This notion of distinct things co-existing in time, yet occupying the same space, is not at all a difficult one. Our consciousness is such a multiplicity of states or qualities all in one. The idea of a group of figures has a very real existence for the sculptor, and he may visualise it with almost all the appearance of reality that the actual, material piece of statuary possesses. In his mind it is a real manifold existence, which nevertheless does not occupy the three-dimensional space which the marble fills. The musical notes C, F, A, C, heard in arpeggio, are things which possess real existence, but which are extended in time, and when we think of these separate sounds we lay them alongside each other in our mind in an empty, homogeneous medium which seems to be all that we think of as space. Yet the same notes heard simultaneously as a chord are not extended. They interpenetrate each other, but yet they are distinct things, since on hearing the chord we can recognise the notes composing it. As an arpeggio the notes are an extensive manifoldness, but as a chord they are an intensive manifoldness.

The mechanistic biology of the latter part of the nineteenth century based itself on the methods and concepts of physics, and it was therefore compelled to assume that the manifoldness of the " primitive " life-substance—the " Biophoridæ " of Weismann and his followers—or that of the fertilised ovum, was a manifoldness that had spatial extension. All the systems studied by physics were aggregates of elements, or parts, that had such extension : the sun, with its attendant planets and satellites, was a system of bodies isolated from each other in space. Even the atmosphere, or the sea, media which to our unaided senses appear to be homogeneous, are really media consisting of discrete bodies, or molecules, which are not actually in contact with each other, but which are separated from each other by empty space. Chemical compounds were assemblages of molecules, molecules were assemblages of atoms, and the atoms themselves were either simple or were composed of corpuscles, or still smaller bodies. This mode of analysis was forced upon the human mind by formal logic and geometry, and it was apparently the only method of acquiring mastery over nature. Yet there were difficulties, appreciated no less by the philosophical physicists than by the writers on formal philosophy. How could bodies, or molecules, or atoms that were separated from each other act upon each other ? The molecule A could only act upon the molecule B if there were some particles between them which could convey the impulse or attraction, but then we must suppose that there were other particles between these intermediate ones, and so on *ad infinitum*, otherwise how could a body act, that is, really exist, where it was not ? In other words, how could there be action at a distance ? How, for instance, could the atoms of the earth attract those of the moon with a force sufficient to break a steel rope of 400 miles in diameter ? Physics had therefore to invent the ether of space, not only to account for interstellar or interplanetary gravitation and other modes of radiant energy, but also to account for the interaction of the atoms or molecules which make up chemical compounds. In our own day atoms have ceased to be the limits to the subdivision of things : they are composed of electrons, but the electrons are entities separated from each other by empty space. They are not, however, the ultimate limits of subdivision of matter, as the atoms were supposed to be by the chemistry of the early part of the last century, but are regarded as " singularities " in an universal continuous medium or ether. It is of no moment that we are unable to describe the ether in terms of our former concepts of matter and energy, or at least that we can only so describe it in such a way that it is represented by negative qualities : we are compelled to postulate its existence in order to avoid philosophical confusion. The universe is therefore a continuum, and an atom or any other body exists wherever it can act. The atoms of a fixed star, so far away that we can only represent its distance in billions of miles, are nevertheless on our earth as well as at the point of space which we regard as their astronomical position, for the light emitted by them acts on our retinas. The universe is an unitary thing in that it is a continuous medium or substance in the philosophic sense, but it is also a multiplicity in that singularities or conditions of this medium pervade each other throughout space. Such seem to be the conclusions towards which the later physics forces us, and it is interesting to reflect how different biological speculation might have been had it been formulated now instead of half a century ago !

Why has a process of evolution occurred at all ? Why is it that tendencies that might have co-existed, that indeed do co-exist to some extent, have become separate from each other ? It is possible to conceive of an organism which contains chlorophyll, and which might therefore synthesise carbohydrate and proteid from inorganic substances, but which might also contain a sensori-motor system, and which might therefore expend the energy so obtained in regulated movements. To a certain extent such organisms combining the plant and animal modes of metabolism do exist among the Protista. Yet, the effect of the evolutionary process has been more and more to dissociate the plant and animal modes of metabolism until the typical animal is quite unable to make use of carbon dioxide and water as materials to be synthesised, while the typical plant has lost all power of motion except the tropistic

U

movements of its roots, leaves, and stems. Instinctive and intelligent behaviour coexist in many animals, yet the tendency of man, most highly intelligent of all, is more and more to act intellectually; while the opposing tendency, that is, to act instinctively, has been evolved in the Hymenoptera. It seems as if such contrasting methods of transforming energy, or of acting, were incompatible with each other, and yet it is clear that they are not really incompatible, for they may co-exist. But it does seem clear that each of these contrasting tendencies cannot be manifested to the fullest extent if it is accompanied by the other. That is to say, life is limited in its power over inert matter. Manifested in the same material constellation, it cannot both use solar radiation to build up substances of high potential energy and then break down these substances so as to obtain kinetic energy of movement. Now we see clearly that life on our earth is indeed limited to a very restricted range of physical conditions. When we think of the mass of the earth we are surprised to find what an insignificant fraction of all this matter displays vital phenomena. The surface of the land is clothed with a layer of vegetation, luxuriant and abundant as we see it when we walk through a tropical forest, but which is really a film of inconceivable tenuity when we compare its thickness with the diameter of the globe. Even the whole surface of the land is not so clothed with vegetation, for polar regions and the tops of high mountains are almost lifeless, while desert tracts may be absolutely so. The lower strata of the atmosphere are inhabited by birds, insects, and bacteria, but the total mass of these is infinitesimal when compared with the total mass of the gases of which the atmosphere is composed. Even the sea, which we regard as rich in life, is not really so : estimates of the luxuriance of planktonic life are really misleading, for although a single drop of water may contain some hundreds of organisms, the mass of these is exceedingly small and is usually expressed as one or two parts per million. All this means that life has difficulty in manifesting itself in material forms. Whether it be simply a mode of interaction of some complex chemical substances with a relatively simple physico-chemical environment —the mechanistic view—or whether it be an impetus or agency which is neither physical nor chemical, but which acts through physical and chemical elements— the vitalistic view,—life is capable of acting on terrestrial materials to a very limited extent. Acting through all the tendencies which we see to exist in it, life may be, so to speak, diluted ; but by being concentrated in one or a few of them it becomes more effective. The dissociation of this bundle of tendencies which we call life is therefore the meaning of the evolutionary process.

Ontogenetic development, says Roux, is the production of a *visible* manifoldness. It cannot be said that this cautious description of the developmental process has been apprehended by those who expound the dogmas of mechanistic biology. Development is indeed the production of a diversity, but this diversity is only a phase of a preceding diversity, a rearrangement of spatially extended pre-existing elements. How else could the developing embryo and its material environment be regarded as a system of physico-chemical elements, capable of study by the methods of experimental and mathematical physics, except by regarding it as a system passing through phases each of which is a necessary consequence of the preceding one, and each of which contained the same elements separated from each other in space ? Let us think of water occupying a vessel at a high temperature and continually cooling.

The states of this system are (1) the gaseous state in which the molecules of the water are moving at a high velocity and are a relatively considerable distance apart, and in which they are incessantly colliding with each other and with the walls of the vessel; (2) the state of the system consisting of the separate phases, liquid water and gaseous steam in contact with it; and (3) the solid phase, in which the molecular motions almost, or quite, cease. Here the progress of the system through its phases leads to physical diversity and then again to physical homogeneity. But the diversity of the different phases is in a sense an apparent one only : any single phase, or at least those which involve the passage of the system from the gaseous to the liquid phases, and *vice versa*, can be represented by van der Waal's general equation,

$RT = (p + \frac{a}{v^2})\ (v\text{-}b)$. Does anything in modern biological

investigation, except, of course, the speculations of non-physical physiologists, suggest that an ontogenetic process can be represented in such a manner ?

Are the arbitrary " stages " of the embryologists— the ovum, blastula, gastrula, etc., phases in a system in the above sense, the only sense in which the process can be regarded as capable of physico-chemical analysis ? What precisely is the embryo at the close of the process of segmentation ? It is an harmonious equipotential system, that is to say, an assemblage of discrete organic parts or cells, each of which has all the potentialities that every one of the others has. *Any* cell in the blastula may become a cell, or a series of such, in *any* part of the gastrula or pluteus larva. This is what the parts are in potentiality, but actually their individual fates are different. The system is an harmonious one, and each of its parts, although able to do whatever any other part can do, yet does one thing only : it becomes an endoderm cell, or an ectodermal cell, or a part of the skeleton, and so on ; what it does depends on its position with regard to the other cells. An extensive manifoldness or diversity is produced, but this was not the consequence of a preceding extensive manifoldness, for in the preceding stage *all the parts of the system were the same.* The manifoldness of the ovum or blastula—that potential manifoldness which became actual in development—must be an intensive manifoldness, and admitting this we must abandon the comparison of the ontogenetic (and, of course, phylogenetic) processes with the phases of a physico-chemical system in process of transformation. *Evolution is the transformation of an intensive into an extensive manifoldness.*

More than this—much more than this—must be the difference between the transforming systems of physics and the evolving systems of biology. There is a quality, or sense, or direction in all naturally occurring inorganic processes which is not like that of organic evolutionary processes. We return now to the consideration of the second law of thermodynamics, for only in this way can we approach the notion of the vital impetus. If an energy-transformation occurs in inorganic nature, that is to say, if anything happens, the transformation occurs or the thing happens because there were diversities in the system in which it occurred. The condition for inorganic happening is that there must have been differences of energy in the different parts of the system : in the most general sense there must have been diversity of the elements. But with the transformation this diversity disappears, or tends to disappear, and it cannot be restored—that is, differences of energy cannot again be established unless by a compensatory energy-transformation ; that is, energy must be expended on the system from without by some external agency. Whatever else physics shows us it shows us an unitary universe, that is, an universe in which anything that happens affects, to some extent, all the other parts. Therefore the diminution of diversities, or energy-differences, is something that cannot be undone, or compensated, for there is nothing without the universe.[1] ˌ Everything that happens in our universe reduces the possibility of further happening. We desire, at the risk of reiteration, that this principle of energetics should be perfectly clear : inorganic happening, of whatever kind it may be, is a case or consequence of the second law of energetics—*is* the second law itself in a sense. All energy-transformations occur because energy-differences are being diminished, because diversities are being abolished. This is the sense, or quality, or direction of inorganic phenomena.

It is not the direction of organic evolution. In the development of the individual organism what we most clearly see is the progressive increase of diversity of the parts. In phylogenetic evolution one, or a few, simple morphological forms of life have become, and are becoming, indefinitely numerous morphological forms. Diversity is continually increasing. If we cling to the mechanistic view of life, we must suppose that the diversity of the fully developed organism, or that of the organic world with all its species, was also the diversity of the fertilised ovum or that of the primitive life-substance in another phase. Then we commit ourselves to all the crudities of modern speculations on heredity.

With this increasing diversity of form there is a

---

[1] It is assumed that the universe is a finite one. If it were infinite the whole discussion becomes meaningless, and we must *give up* this and other problems.

concomitant segregation of energy. We see as clearly as possible that the tendency of all inorganic happening is the transformation of potential into kinetic energy, and the equal distribution of this kinetic energy throughout all the parts of the system in which the happening occurred. On the other hand, the tendency of organic happening is the transformation of kinetic energy into potential energy, (1) in the stores of chemical compounds which result from the metabolism of the green plants, and which are capable of yielding energy again ; and (2) in the results of the instinctive or intelligent activities of the animal's organism. The first result of organic evolution is clearly to be traced and needs no further explanation, the second is apparent on reflection, but is perhaps not clearly apprehended in all its significance by the student of biology and physics.

Organic evolution is the process which has had, or is having, for its tendency the development of the putrefactive and fermentation bacteria, the chlorophyllian organisms, the Arthropods, and man and other mammals. All that we have said has been futile if this teleological description of the evolutionary process has not been clearly suggested. The indefinitely numerous forms of life that have appeared on the earth in the past, and are now appearing, seem to be experiments most of which have been unsuccessful. Only in the organisms mentioned, organisms which are complementary in their metabolic activities, has life been successful in manifesting itself in activities which are compensatory to those of inorganic nature. The energy which is dissipated in the radiation of the cooling sun is again made potential in the form of the carbohydrates, synthesised from water and carbon dioxide by the agency of the chlorophyllian organisms, and this energy accumulates. It is employed by the instinctive and intelligent animal, in that it is used as food and converted into bodily energy, which can then be utilised for any purpose that is contemplated. These plant substances taken in by the animal as sources of energy are broken down into excretory substances, which are further broken down by the metabolic activity of the fermentation and putrefaction bacteria, and become the substances used as foods by the chlorophyllian organisms.

If the activities of man were only those of undirected or misapplied muscular movements (as indeed most of his activities have so far been), then cosmic energy would truly be dissipated after it had become the energy of organisms. But does not all the history of man point to his ever-increasing activity in the conquest over nature, that is, the effort to hoard and employ natural sources of energy, and to arrest its tendency towards dissipation ?

It must be admitted that the past history of human civilisation has been almost entirely that of the irresponsible exploitation of natural resources—for it has been founded on the thoughtless and wasteful utilisation of energy which was made potential by the plant and animal organisms of the past. Man, the hunter, maintained himself and multiplied by the destruction of other animals or plants, or by the mere collection and utilisation of naturally occurring fruits and other plant-substances. During historic times the bison and other animals have almost become extinct owing to his ruthless activity, just as in our own days the whale, sole, and turbot are disappearing before the activity of the machine-aided fisherman. Industrial man has been successful with his factories and railroads and steamships, and his electrical power and transport, only because he has been able to utilise the stores of energy contained in the coal and oil accumulated in the rocks of the earth. The progress of civilisation has been a progress rendered possible by discovery and invention, and by the application of the knowledge so obtained to the practical things of human life, but in this speculation and its application two different things are indicated. For the scientific man and the philosopher the reduction of the apparent chaos of nature to law and regularity is the beginning and end of his mental activity; but the object of the "entrepreneur" or "organiser" or the "captain of industry" has been to employ these results of thought to the irresponsible exploitation and the selfish depletion of natural sources of energy. Just as the bison and other animals have disappeared or are disappearing before the hunter and fisherman, so the stores of coal and oil are disappearing before the activities of commerce. It has been said that the triumphs of industrialism are only the triumphs of the scientific childhood of our race. Human effort has so far only contributed to the general dissipation of natural energy.

Yet just as man, the hunter, has been succeeded by man, the agriculturalist, so this irresponsible depletion of natural wealth must be succeeded by the endeavour to retard, and not to accelerate, the degradation of energy. Plants and animals which were simply killed by primitive man are now sown and harvested, or cultivated and bred; so that the energy of solar radiation, which formerly ran to waste, so to speak, is now being fixed by the metabolic activity of the green plants of our crops and harvests. Rainfall and winds, tides and rivers, all represent energy primarily derived from solar radiation and from the orbital and rotatory motions of the earth and moon. This energy even now is almost entirely dissipated as waste, irrecoverable, low-temperature heat ; but more and more as our stores of coal and oil are being depleted, the attention of men is being directed to these sources of kinetic energy. Waterwheels and windmills, and the more effective mechanisms that must be evolved from these primitive motors, will capture this waste energy and convert it into the kinetic energy of machines serviceable to man, or into the potential energy of chemical compounds capable of storage and future utilisation. The study of radio-activity has made us acquainted with the enormous stores of potential energy locked up in the atoms, and if it ever should become possible to utilise this by the disintegration of these particles, the downward trend of natural energetic processes will further be retarded.

Life, when we regard it from the point of view of energetics, appears therefore as a tendency which is opposed to that which we see to be characteristic of inorganic processes. The direction of the latter is towards the conversion of potential into kinetic energy, and the equal distribution of the latter throughout all the parts of the universe. The direction of the tendency which we call life is towards the conversion of kinetic into potential energy, or towards the establishment and maintenance of differences of kinetic energy, whereby the latter remains available for the performance of work. In general terms, the effect of the movement which we call inorganic is towards the abolition of diversities, while that which we call life is towards the maintenance of diversities. They are movements which are opposite in their direction.

What is cosmic evolution ? In all the hypotheses which astronomical physics has imagined we see the transformation of a system—a part of the universe arbitrarily detached from all the rest—through a series of stages, each phase of the series being marked by a progressive decrease of diversity, that is, by some degradation of energy.   Two main series of hypotheses accounting for the present condition of the universe seem to have been the result of physical investigation : (1) the origin of discrete solar and planetary bodies by a process of condensation of a gaseous nebular substance ; and (2) the origin of the same systems by aggregations of meteoric dust.   Plausible as is the nebular hypothesis on first consideration, it fails when it is subjected to minute analysis.   What is a gaseous nebula ?   It is a mass of heated vapour contracting by the mutual gravity of its parts as its molecules lose their heat by radiation—so the hypothesis states. But it has been pointed out that we cannot be certain that the gaseous nebulæ known to astronomy are hot, or even that they gravitate.   The great nebula in Orion, it is stated, is at an enormous distance from us, and making a minimal estimate of this distance the volume of the nebula must still be incredibly great.   There are good reasons for believing that the mass of the visible universe cannot be greater than that of a thousand million of suns such as our own.   Assuming that all this matter is contained in the great nebula in Orion (and obviously only a small portion of it can be so contained), we find on calculation that the " gas " so formed would be much less dense than even the trace of gas contained in a high vacuum artificially produced.[1]   How, then, can we speak of such a body as this nebula as an extended mass of hot gas, cooling and gravitating as it loses heat ?

Even on the other hypotheses, those of the formation of discrete suns and planets by the aggregation of meteoric dust, and the compensatory dispersal of such

---

[1] Its density would be $\dfrac{1}{58 \times 10^8}$th that of our atmosphere.

dust by radiation pressure, apparently insurmountable difficulties arise. All such hypotheses as we have indicated assume material substance and modes of energy-transformation similar to those that we study in laboratory processes, and all such hypotheses involve the notion of the degradation of energy. So long as we suppose that all cosmic processes are transformations of extended systems of material substances we must assume that energy is dissipated at every stage of the transformation, and whenever we assume this we admit that the processes are irreversible ones, and that the material universe as a whole tends towards a condition of inertia. Yet this, we see, cannot be true, for the universe teems with diversity. Is the progress towards the ultimate state of inertia an asymptotic one, as Ward suggests? This does not help us, since all that the suggestion does is to misapply a mathematical device of service only in the treatment of the problems for which it was developed. Somewhere or other, it has been said, the second law of thermodynamics *must* be evaded in our universe.

How can it be evaded? That movement or progress which we call inorganic is a movement of energy-transformations in one direction—towards their cessation. It is a movement which we can easily reverse in imagination. A cigarette consumed by a smoker represents the downfall of energy : the cellulose and oils of the tobacco burn with the liberation of heat, and the formation of water, carbon dioxide, and some soot ; and this is what happens when potential energy contained in an organised substance becomes converted into kinetic energy. Now, the opposite process can clearly be conceived—it can even be pictured. If we make a kinematographic record of the smoking of the cigarette and then reverse the direction of motion of the film, we shall see the particles of soot recombining to form the substance of the cigarette, and we can imagine the concomitant combination of the water, carbon dioxide, and other substances formed during the combustion with the absorption of kinetic energy. This is not a mere analogy, for the same reversal of ordinary chemical happening occurs whenever a green plant builds up starch from the water and carbon dioxide of the atmosphere ; and it also occurs whenever a chemical synthesis of an " organic " compound, like that of urea by Wöhler, or that of the sugars by Fischer, is brought about in the laboratory. In all such syntheses the experimenter *reverses* the direction of inorganic chemical happening. He may cause endothermic chemical reactions, reactions accompanied by the absorption of available energy, to take place, and in these kinetic energy becomes transformed into potential energy. All the syntheses of organic compounds so complacently instanced by mechanistic biologists and chemists as indicative of the lack of distinction between the organic and the inorganic point to no such conclusion. Sugar is built up in the cells of the green plant from the inorganic compounds, water, and carbon dioxide, and is therefore a compound prepared by life—that of the plant organism. But sugar may also be built up in the laboratory from inorganic compounds, which may further have been synthesised by the chemist from their elements. Does this destroy the distinction between compounds formed by the agency of the organism and those formed by inorganic agencies ? Obviously it does not, for in the green plant the sugar was formed as the result of the vital agency of the living chlorophyllian cell, while in the laboratory it was built up because of the intelligence of the experimenter. Apart from this intelligence or vital agency, the series of chemical transformations beginning with the elements carbon, oxygen, and hydrogen, and ending with the substance sugar, would not have occurred. We have no right to say, therefore, that such syntheses destroy the distinction between the organic and the inorganic. What they do indicate is the distinction between the tendency expressed by the second law of thermo-dynamics (inorganic processes), and those that occur as the result of direction conferred upon processes taken as a whole, either by the vital agency of the living cell, or by the intelligence of man (vital processes).

The direction, therefore, that may be conferred on a series of physico-chemical processes is what we must understand by the " vital impetus " of Bergson, or the " entelechy " of Driesch.

It must be admitted that it is difficult to describe more precisely than we have done above what is meant by these terms. It is with very much the same em-barrassment that is experienced by the physicist when he has to apply the concepts of mass and inertia, in their eighteenth-century meaning, to his·description of an universe in terms of electro-magnetic theory, that we seek to describe the modern concept of entelechy. Yet the physicist has had to make this step forward, and the same adventure awaits the biologist if the speculative side of his science is to make further pro-gress, and if he is disinclined to make his science an appendage of physics and chemistry. Entelechy does not correspond to the eighteenth-century notion of a " vital force," or to the " soul " of Descartes, as the writer of a book on evolutionary biology seems to suggest. It is a concept which is forced upon us mainly because of the failure of mechanistic hypotheses of the organism. If our physical analysis of the behaviour of the developing embryo, or the evolving race or stock, or the activities of the organism in the midst of an ever-changing environment, or even the reactions of the functioning gland, fail, then we seem to be forced to postulate an elemental agency in nature manifesting itself in the phenomena of the organism, but not in those of inorganic nature. This argument *per ignorantium* possesses little force to many minds : it makes little appeal to the thinker, or the critic, or the general reader, but it is almost impossible to over-estimate the appeal which it makes to the investigator, as his experience of the phenomena of the organism increases, and as he feels more and more the difficulty of describing in terms of the concepts of physics the activities of the living animal.

We may, however, attempt to illustrate mainly by analogy what is meant by Driesch's *entelechia*, a more precise concept than is Bergson's *élan vital*. We return to the consideration of the behaviour of the embryo at the close of the process of segmentation. The organism at this stage consists of a number of cells organically in continuity with each other, either by actual proto-plasmic filaments or by the apposition of parts of their surfaces, thus constituting " semi-permeable " mem-branes. These cells are all similar to each other, both structurally and functionally. It does not matter that modern speculations on heredity describe them as unlike in that each contains a different part of the original germ-plasm which had been disintegrated in the process of the division of the ovum and the first few blastomeres ; and it does not matter that these hypotheses are compelled to assume that a part of the original germ-plasm remains intact, being destined to form the gonads of the adult animal. These are hypo-theses invented to account for the differentiation of the embryo in terms of eighteenth-century physics and chemistry, and they have yet to be supported by experiment before we can accept them as a *description* of what is to be observed in the processes of nuclear division and segmentation. Further, it is certainly the case that any one cell of the early embryo can give rise to any part of the larva. The segmented embryo is therefore a system of parts, all of which are potentially similar to each other. But actually each of these parts has a different fate in the process of the development of the larva, and this fate depends on what is the fate of the adjacent cells. There is also a plan or design in the development of the embryo—that is, a very definite structure results from this process—and each of the cells shares in the evolution of this design. The system of cells is therefore an harmonious equipotential system. The cells themselves are not the ultimate parts of this system, for each of them is an aggregate of a very great number of substances which are physico-chemically characterised—at least our methods of analysis seem to show that each cell is a mixture of a number of chemical compounds, but we must never forget that it is the dead cell which we thus subject to analysis, and not a living organism. Let us call these supposed chemical constituents of the living cells the elements of the system ; then at the beginning of the process of development the latter is composed of elements which are not definitely arranged but which are distributed in an " homogeneous " manner very like the distribution which is effected on shuffling a pack of cards. But as differentiation proceeds, the elements of this system become unequally distributed, and the diversity becomes greater and greater, attaining its maximum when the definitive tissues and organs of the adult become established, just as at the close of a game of bridge the cards acquire a particular arrangement indicative of a very definite plan which was present in the minds of the players shortly after the game began.

Mechanistic biology would seek to explain this transformation of a homogeneous system of elements into a heterogeneous and specific arrangement by the interaction of the elements with each other, and by the reaction of the environment. But, given a homogeneous arrangement of elements capable of interacting with each other, then only one final phase can be supposed to be produced. A mixture of sulphur, carbon dust, copper and iron filings raised suddenly to a high temperature will only interact in one way, and the final phase of the system will depend on the composition of the mixture, on the temperature, and on the conduction of heat into the mixture in the initial stage of heating. A mixture of chloroform and water shaken up in a bottle is at first a " homogeneous " mixture of the particles of the two substances, but under the influence of gravity the liquids separate from each other and form two distinct layers, each of which will contain in solution some of the other liquid. A homogeneous mixture of different substances therefore becomes a heterogeneous arrangement in the inorganic system, as in the organic one, but while we can predict the former one we cannot predict the latter. We can express the result of the combination of the elements of the inorganic mixture as something that depends on chemical and physical potentials, but this is quite impossible in the case of the development of the embryonic system. It is not only that our knowledge of the developmental process is imperfect : the distinction between the two processes of differentiation is a fundamental one. A change in the conditions under which the inorganic system differentiates leads

x

of necessity to a different final phase, but a change in the conditions under which the embryo develops need have no such effect. If some unforeseen occurrence takes place—some artificial interference with the process of segmentation, which could never have been experienced in the racial history of the organism—a *regulation* by the parts of the embryo occurs, and the final phase of development may be the same as if no interference had been experienced. That which is operating in the development of the embryo is something that is permitting, or suspending, or arranging physico-chemical reactions.

Let us think of the developing embryo merely as an aggregation of substances contained in an inorganic medium : the segmented frog's egg floating on the water at the surface of a pond is an example. As an inorganic system its fate is determined. Autolysis of the substances in the cells will occur and the proteids will break down with the formation of amido-bodies, while other chemical changes, strictly predictable if our knowledge of organic chemistry were more complete than it is, would also occur.   Putrefactive and fermentative bacteria will attack the proteids, fats, and carbohydrates, and in the end our aggregation of chemical substances will become an aggregation of much simpler compounds—water, carbon dioxide, marsh gas, sulphuretted hydrogen, phosphoretted hydrogen, ammonia, nitrates, etc., all of which will dissolve in the water of the pond, or will diffuse into the adjacent atmosphere. But in the living embryo this is not what occurs : an entirely different, and much more complex, arrangement of the chemical substances originally present in the segmented egg, or at least a physical and chemical re-arrangement, is brought about. The entelechy of the developing embryo prevents some reactions from occurring and directs the energy which is potential in the system towards the performance of other reactions.

Two analogies, suggested by Driesch, will perhaps make the rôle of entelechy more clear. A workman, a heap of bricks, some mortar, some food, and some oxygen constitute a system in the physico-chemical sense. From his heap of bricks and mortar the workman may build one of several different kinds of small house, or he may perhaps construct several walls without any definite arrangement, or he may merely convert one " disorderly " heap of bricks and mortar into another " disorderly " heap. In the same way a man, a case of movable types, some food, and some oxygen constitute another system. The initial phase of this system consists of the compositor, his food, and some fifty-odd boxes of types, each of which contains a large number of similar elements. A final phase of the system may be the arrangement of the types to form an epic poem, or a series of dramatic criticisms, or a meaningless jumble of correctly spelt words. In all these cases the same amount of energy was expended : the bricklayer used up the same quantity of food and oxygen and excreted the same quantities of water, carbon dioxide, and urea, whether he made a house, or a small chimney, or a heap of bricks without architectural arrangement. The system of bricks and mortar acquired during the process of differentiation a gradually increasing complexity ; while in the case of the type-setting the diversity of arrangement acquired in the final phases may be of a very high order. Yet the intelligent mind of the worker remained in either case unchanged.

Let us consider further a man walking along the ties, or sleepers, of a railway track. The ties are at variable distances apart, so that the steps of the walker must vary in length, being sometimes closer together, sometimes further apart. The *mean* step has a definite length and requires the expenditure of a certain amount of energy, and the condition that the man takes sometimes a long step and sometimes a short one does not require that the energy expended on the steps should be more than if every one of them were of the mean length, for the additional energy that is required for the long steps is saved from the short ones. That which operates here is the power of regulation exercised by the walker regarded as a mechanism. There is no purely inorganic process precisely similar to this. It might be thought that the governor of a steam engine did very much the same thing, admitting more steam into the cylinder when the load on the engine increases, and *vice versa*. But the governor is a mechanism *designed* to compensate for variations *that are given in advance*. In the case of the man walking on the railway track, entelechy operates by suspending energetic happening (the muscular contractions of the short steps) when necessary, and allowing it to proceed when necessary. Entelechy itself, whatever it may be, need not be affected by these regulations.

The organism is therefore an aggregation of chemical substances arranged in a typical manner. These substances possess energy in the potential form, capable of undergoing transformation so that they may give rise to other chemical substances—secretions, for instance—or to energy in the kinetic form, that is, the movements of muscles. In the resting organism these transformations do not take place : the energy remains potential, so that chemical happening is suspended. In the unfertilised ovum, for instance, nothing happens although all the potentialities of segmentation are contained in the cell. If reactions did occur in consequence of the chemical potentials contained in the substances of the cells, the progress of these would be such as to lead to the formation of substances in which potential energy was minimal, and in which the original energy of the cell would be represented by the un-co-ordinated kinetic energy of the molecules resulting from the breakdown of the substances undergoing the chemical changes. This is not what happens in the differentiation of the ovum : the developing cell forms new substances from those of its inorganic medium similar to the substances of which it is already composed, and then these substances become arranged to produce the specific form of the organism into which the ovum is about to develop.

All hypotheses which attempt to describe the functioning of the differentiating ovum, or the functioning organism, in terms of the physical concepts of matter and energy alone, fail on being subjected to close analysis. The manifestations of the life of the organism are, it is said, particular " energy-forms," of the same order as light, heat, chemical and electrical energy, etc. All these energy-forms are " concatenated," that is, each can be converted into any of the others. A particular frequency of the vibration of the ether can be converted into a movement of the molecules of a material body, and so become heat, while chemical energy may become converted into electrical energy, or *vice versa*, and so on. It is said that life may be merely a transformation of some " energy-form " known to us : the potential energy of food may be converted into " biotic energy," and this may then manifest itself in the characteristic behaviour of the organism. This is the method of physical science. Energy continually disappears from our knowledge : the mechanical energy which was employed to carry a weight to the top of a hill, or that which raises a pendulum to the highest point of its swing, apparently disappears. If we pass a current of electricity through water, energy disappears, for it requires more current to pass through water than through a piece of metal of the same section. In these and similar cases physics invents potential energies in order to preserve the validity of the law of conservation. The kinetic energy of the weight, or that of the swinging pendulum, becomes the potential energy of the weight resting at the top of the hill, or that of the bob of the pendulum at its highest point, while the electrical energy that has apparently been lost becomes the potential energy of the changed positions of the molecules of oxygen and hydrogen. This assumption that the visible kinetic energy of motion becomes converted into the invisible potential energy of position is justified by our experience, for (neglecting dissipation) we can recover this lost energy, in its original quantity, from the condition of the bodies which became changed physically when the kinetic energy disappeared. Apply the same method to the phenomena of the organism and suppose that the chemical potential energy of the food consumed becomes converted into the kinetic energy of motion of the parts of the body : we are justified in this assumption by the results of physiology. But then some of this chemical energy undergoes a transformation of quite another kind and becomes the "biotic energy," which is apparently that which is in us which enables us to perform regulations, or establishes that condition which we call consciousness. We cannot say exactly what this "biotic energy" is, or what are the steps by which the energy of food becomes converted into it ; but no more can we say what is electrical energy, nor what are the steps by which chemical energy becomes converted into it. Thus our ignorance of the precise nature of the energy-transformations of inorganic things—an ignorance which is all the while disappearing—becomes the excuse for a comparison of these with vital transformations, and for the assumption that there is a fundamental similarity in the two kinds of happening.

Less is assumed in the assumption of an entelechian agency than in assuming that the manifestations of life are the consequences of a vital " energy-form," different from inorganic forms, though belonging to the same order, inasmuch as it may be concatenated with these inorganic energy-forms. We need not suppose that a particular kind of transformation occurs only in the sphere of the organic : all that we need assume is that, by some agency inherent in the activities of the organism, chemical reactions that would occur if the constellation of parts were an inorganic one are suspended. Nothing unfamiliar to physical science is involved in this assumption. Hydrogen and chlorine, gases that combine together when mixed with the production of heat and light, may be mixed under conditions such that the combination may be delayed for an indefinite time. Iron which dissolves in nitric acid may nevertheless be brought into the " passive " form when it remains in contact with the re-agent but is not dissolved by it. Enzymes which are in contact with the walls of the alimentary canal do not dissolve these membranes so long as the tissues are alive, and they do not dissolve the food stuff until they have been " activated." Oxygen which is contained in the tissues does not oxidise the tissue substances until an enzyme or a catalase has exerted its influence. More and more, as physiology has become more searching in its study of the functions of the animal, has it sought to explain the metabolic processes by assuming the intervention of enzymes, until the number of these substances has become legion, and much of the original simplicity of the notion of ferment-activity has been lost. But why do not these enzymes, if they are always present in the tissues, always act? They must be activated, says modern physiology; that is, the enzyme really exists in the tissues as a " zymogen " or a substance which is not, but which may become, an enzyme; or they exist as " zymoids," that is, substances which appear to be chemically enzymes, but which must be activated by "kinases" before they can become functional.

Undoubtedly it is along these lines that physiology is making advances, has increased our knowledge of the *activities* of the animal, and is conferring on the physician greater power of combating disease ; but the hypotheses of the activity of the enzymes is obviously one which has been based on the results of the physico-chemical investigation of inorganic reactions, and it has taken the precise form it has because of the attempted analogy of many metabolic processes with catalytic processes. Why do the inert zymoids become activated by the kinases just when they are required by the general economy of the whole organism? We do know that kinases are produced by the entrance of digested food into certain parts of the alimentary canal, and that these kinases are carried in the blood stream to other parts where they activate the zymoids already there. But of the nature of the machinery by means of which all this is effected physiology gives us no hint, and it is an assumption that the mechanism involved is a purely physico-chemical one. Suppose we say that the entelechy of the organism possesses the power of suspending the activation of the enzyme, that is to say, of arresting the drop of chemical potential involved in the process of the hydrolysis of (say) a proteid. When this process of hydrolysis is necessary in the interest of the organism entelechy can then institute the reaction which it has itself suspended : all this is in accord with the law of conservation. Entelechy does not cause chemical reactions to occur which are " impossible " : it could not, for instance, cause sulphuric acid and an alkaline phosphate to react with the formation of hydrochloric acid. But chemical reactions which are possible may be suspended, and suspended reactions may then become actual when this is necessary in the interest of the organism.

Entelechy is therefore not energy, nor any particular form of energy-transformation, and in its operations energy is neither used nor dissipated. In all that it does the law of conservation holds with all the rigidity with which we imagine it to hold in purely inorganic happening—at least we need not assume that it does not hold—and this is the essential difference between the entelechian manifestations and the manifestations of the " vital " or " biotic " forces or energies of the historic systems of vitalism. It is essentially arrangement, or order of happening, and it is therefore a non-energetic agency. The workman who may build half-a-dozen zigzag walls, or an archway, or a small house, from the same materials and with the expenditure of the same quantity of energy, is indeed an energetical agent, but he is more than that. He is a physico-chemical system in which any one phase is not determined by the preceding phase. Different results may arise from the same initial arrangement of materials and energies, and this is because the system contains more than the material and energetical elements. It contains the intelligence or entelechy of the workman.

What is this entelechy? Sooner or later in all our speculation on organic happening we must cross the arbitrary line which divides the space of our concepts from the non-spatial—the intensive from the extensive. Just as the physicists have left materiality behind them in their speculations and treatment of the phenomena of radiation, so biology must attempt to trace back the materiality of the organism to something which is immaterial. Just as physics has now abandoned the idea of matter as something which consists of discrete particles, or atoms, having extension in space, and which therefore exclude each other, so biology must seek the origin of living things, not in the hypothetical "biophoridæ," or other ultimate living material particles, but in the intensive manifoldness of entelechy. There *is* a manifoldness in the potentiality which the simple and homogeneous ovum possesses of becoming the heterogeneous adult organism. This manifoldness, says the mechanistic biologist, consists of a manifoldness of extended material units, the determinants of Weismann, and the organisation that arranges these units—what is this organisation? It cannot be a three-dimensional machinery, as all close analysis of the facts of development and regulation shows. It is then something that is intensive, something which is not in space, but which *acts into space*, and the result of which is manifested in spatial material arrangements and activities. Vague and incomprehensible as is this concept of the activities of the organisms, it is only vague and incomprehensible because we have been accustomed to express all chemical and physical happening in terms of the fundamental concepts of matter and energy, and the science of the last two centuries has left us with a terminology which applies strictly to operations in which only these concepts are involved. But if, as all minute analysis of vital phenomena shows, the search for the antecedents of some energetic, material, extended system of elements in a preceding energetical, material, extended system of elements only leads to confusion and contradictions, then this concept of an agency which is neither energetic, nor material, nor spatial must be formulated. Entelechy, then, is not energy, but rather the arrangement and co-ordination of energetic processes. It is not something that is extended in space, but something which acts into space. It is not material, but it manifests itself in material changes. It is a manifoldness, or organisation, but the manifoldness is an intensive one. Compare this definition with the notion of the ether of space now accepted by the mathematical physicists, and it will be seen that our speculations are similar to those of the physicists, and, like them, the test of their reality and usefulness is to be justified pragmatically.

We may now attempt a formal description of the organism based on the discussions of the previous chapters.[1]

The organism is a typical constellation of physico-chemical parts or elements.

That is to say, it is an object in nature possessing a definite form, which is the result of the arrangement of its tissues. Each tissue is again an arrangement of cells, and each cell is a complex of chemical substances. The organism therefore resembles, so far as our definition goes, an inorganic crystal. But it is the typical

[1] This description is largely an expansion of Driesch's "Analytical definition of the individual living organism." The reader should note also that it includes the Bergsonian idea of duration, and that of the organism as a typical phase in an evolutionary flux, as parts of the description.

organism that we are considering, and this is a pure conception, for our typical organism does not occur in nature. The organisms that are accessible to our observation are constellations of physico-chemical parts, but these constellations tend continually to deviate from the conceptual arrangement. Progressive variation from the type is something that distinguishes the organic constellation from the inorganic one.

The organism is an entity in which energy-transformations of a particular nature are effected. These transformations raise energy from a state of low, to a state of high potential.

This is the general tendency of terrestrial life, and it is expressed most fully in the metabolism of the green plant. The energy-transformations that are effected here are those in which the kinetic energy of radiation is employed to build up chemical compounds of high potential, from inorganic substances incapable in themselves of undergoing further transformations. The general tendency of all inorganic transformations is towards inertia. In them energy is not destroyed, but it is dissipated : it becomes uniformly distributed throughout material bodies as the un-co-ordinated motions of the molecules of which those bodies are composed, and it ceases to be available for further transformations. The green plant reverses this transformation, and accumulates energy in the form of chemical compounds of high potential. Inorganic processes are those in which available energy becomes unavailable, and this unavailable energy can only become available again if a compensatory energy-transformation is effected. Life is that which effects these compensatory energy-transformations.

The organism is a constellation capable of indefinite growth by dissociation.

That is to say, it is a constellation which reproduces itself in all its specificity. Growth consists in the separation from the organism of a part, or reproductive cell, which divides (or dissociates) repeatedly, each dissociated part growing again in mass by the addition of substances similar to its own, but which are taken from a medium dissimilar in composition to itself. The aggregate of parts so formed then differentiates so that the constellation is reproduced in all its specificity. There is nothing precisely similar to this in inorganic happening. The growth of a crystal consists simply of the accretion of elements similar in nature to those of the growing body, and there is no differentiation.

The organism exhibits autonomy.

It is a constellation which persists in the midst of an ever-changing environment, and the typical organic form remains the same, although the material of which it is composed undergoes continual change. There are inorganic entities which resemble the organism in this respect : the form of a cyclone or atmospheric disturbance, for instance, remains the same even though the air of which it is composed is continually changed. But the form of the organism does not vary strictly with the changes in the environment in which it is placed, for it may respond to an environmental change by a regulation, or compensatory change in form or functioning, the effect of which is to maintain the constellation in all its specificity. The regulation is not a complete or perfect one, for environmental changes do, to some extent, produce changes in the organic constellation, but there is no functionality between the environmental change and the organic response. In inorganic happening a change in one part of a transforming system necessarily determines the nature and extent of the changes that occur in the other parts of the system.

The organism is a centre of continuous action.

It is first of all a part of nature in which energy-transformations continually take place—a description which applies equally well to plants and animals. It is only when we attempt to seek an inorganic system to which this definition would apply that we find how well it differentiates the organic from the inorganic. An inorganic system which transforms energy is either one which tends continually towards stability, or it is a machine made by man for a definite purpose, and it is therefore a system involving a teleological idea. An organic centre of action is one in which energy-transformations proceed without cessation.

In the plant organism the energy-transformations represent, with the exception of the reproductive processes, the whole activity of the organism. In the animal organism they are accessory to regulated and purposeful motile activity, that is, muscular action. The object of this muscular activity varies with the stage of evolution attained by the animal. Its sole object in the lower animal is that of individual or racial preservation. Living in an organic and inorganic environment which is always hostile and tends continually towards its destruction, the whole activity of the organism is directed to the attempt to master this environment : it struggles for its individual existence, and that of its offspring. The activities of man are also these, but they are more than these, for, knowing that physical processes tend continually towards inertia, he seeks to control these processes, and to preserve the instability of nature on which the possibility of further becoming depends.

The activity of the organism, whether it be the energy-transformations of the plant or the motile activities of the animal, are directed and regulated activities. The activity of the organism is not a functional activity in the sense that the activity of a dynamo is a function of the nature of the machine, and of the nature and quantity of the energy supplied to it. The nature of the activity of the organism is regulated autonomously by purposes which it " wills " to carry out.

The organism is a phase in an evolutionary flux.

Categories of organisms—varieties, species, genera, etc.—are fictions. They are arbitrary definitions designed to facilitate our description of nature. They are types or ideas. In constructing them we follow the method of the intellect, and we represent by immobility that which is essentially mobile and flows. Between the fertilised egg and the senile organism there is absolute continuity. Our description of the individual organism is a description of it at a typical moment of its life-history, and this description includes all that has led up to, as well as all that will fall away from, the morphology at this particular typical moment.

Even then the arbitrarily defined organism is only a phase. In defining it we arrest, not only the individual, but also the racial, evolutionary flux. The specific morphology is that of a typical moment in a racial flux. Leading up to it at this moment are all the variations that have joined it with its ancestry, and leading away from it will be all the variations that will convert it into its descendants.

The individual and racial developments are true *evolutions*. They are the unfolding of an organisation which was not expressed in a system of material particles or elements interacting with each other, and with the elements of the environment, but which we must seek in an intensive, non-spatial manifoldness.

In the evolutionary flux the changes are non-functional ones, that is to say, any phase, whether it be one in an individual or a racial development, is not merely a rearrangement of the elements of the preceding phases, as in the case of a transforming system of material particles and energies. There is inherent, spontaneous variability.

The organism endures.

That is, all its activities persist and become part of its organisation. It does not matter whether or not we decide that characters which are acquired are transmitted, nor does it matter whether or not we conclude that the environment is the cause of these acquirements. Some time or other in the individual or racial history new characters arise by the activity of the organism itself, and these characters either persist in an individual or in a race. They endure. All its activities, even its thoughts, persist and form the experience of the animal—an experience which continually modifies its conduct. In man those true acquirements, the results of education and of investigation, persist as written language, or as tradition, even if they are not inherited.

Duration is not time. The mathematician does not employ, in his investigations, intervals of duration. When he relates something which is happening now to something which happened some time ago he employs the differential co-efficient dy/dx, so that the interval between the two occurrences becomes an " infinitesimal " one. When the astronomer predicts events that will happen some years hence, or describes those that happened some years ago, he is really describing things that are all there at once, so to speak, things which are given. If we unfold a fan, stick by stick, we see the separate members in succession, but they are all there, and we can, if we like, see them all at once.

The more we reflect on it the more we see that mathematical time is only a way in which we see things apart from each other. Things become extended in time as they become extended in space. Whether occurrences capable of analysis by the methods of physics are what we call past or future occurrences, they are all given, in that each of them is only a phase of the others.

Duration belongs to the organism. The past is known because all that has occurred to the organism still persists in its organisation. The future is unknown because it has still to be made. Duration is therefore a vector—something having direction, and the organism progresses out of the past into the future. It grows older but not younger.

Such appears to be the nature of life. Can we discuss the problem of its *origin* ?

Did life originate on our earth ? We must first consider what we mean when we speak of an origin. The organic world of the present moment, with all its environment—that is to say, the totality of organisms on the earth, with all the materials which they can utilise in any way, the energy of radiation from which they ultimately derive their energy, and all the parts of the cosmos which interact with them—constitute a system in the physical sense. The present condition of the organic world, that is, the kinds and numbers of organisms, and their distribution, and the distribution of the materials which they can utilise, and the quantity and nature of the energy which is available to them, are the present phase of this system. All the conditions of life in the past, that is to say, the kinds, and numbers,

Y

and distribution of organisms, and the quantity and nature of their environment at any time, together formed phases of this system. If there was a time when life, as we know it, did not exist, then the materials and the energies, which were antecedent to life when it did appear, were also a phase of the system. On a strictly mechanistic hypothesis there could be no origin : there could only be a transformation of a system which was already in existence. All that exists to-day was given then. When, therefore, we speak of the origin of life from non-living materials we mean simply a transformation of those materials and energies.

There was a time, it is said, when life could not exist on the earth. For the organism is essentially that aggregate of chemical compounds which we call protoplasm, and this cannot exist at temperatures higher than 100° C., and it cannot function at temperatures lower than 0° C. It requires carbon dioxide, and ammonia or nitrate, as the materials for its constructive metabolism, and there was a time when these compounds could not exist, for they must have been dissociated by the heat of the gaseous nebula from which our earth originated. The organism requires energy in the form of solar radiation of a particular frequency of vibration, and there was a time when the sun's radiation was different from what it is now. Therefore life did not exist then. Even if we believe that life came to the earth as germs, which existed previously in outer cosmic space, this belief does not solve the problem, which simply becomes transferred from our earth to some other cosmic body.

But life, as we know it, makes use of the materials and the energies which are available to it in the conditions in which it exists. The plant organism obtains its energy from solar radiation because this is the most abundant source of terrestrial energy. The human eye is most susceptible to light of a particular frequency of wave-length, but this is the radiation that is most abundant in the light of the sun. Does this not mean that the organism has merely adapted itself to the material and energetic conditions in which it exists ? Does it necessarily mean that because the conditions were very different life could not exist ? Protoplasm could not exist at a temperature of several thousand degrees Centigrade, but does that mean that life, which on any hypothesis of mechanism must be described in terms of energy, could not exist in these conditions ?

It must have had an origin, says Weismann, because it has an end. Organic things are destroyed, inasmuch as they disintegrate into inorganic things. Organisms die. Thus the organic process comes to an end, and because it comes to an end it must have a beginning. Spontaneous generation of life is thus, for Weismann, a " logical necessity."

Need this logical necessity exist ? The argument clearly implies that life is a reversible process. Organic things become inorganic, and therefore inorganic things must become organic things. The first statement is a fact of our experience, but the second one would only be logically true if we were to postulate that the process of life, whatever it may be, is a reversible process. But we must not postulate this if we are to hold to a physico-chemical mechanism, for it is a fundamental result of physical investigation that all inorganic processes are irreversible : reversible inorganic processes are only the limits to irreversible ones. Physical processes go only in one way, and that organic substance is destroyed to the extent that it becomes inorganic is a particular case of this irreversible physical tendency. Now the mechanism of Weismann must base itself on the concepts of physics and chemistry, and it must postulate the origin of life from non-living substances. Why ? Because life is a reversible process, that is, it exhibits a tendency which does not exist in inorganic processes. Clearly the logic is faulty ! And must we conclude that life has an end ? Weismann himself suggests that nothing in the results of biology indicates that physical death is a necessity : it is rather an adaptation. The soma, or body, is the envelope of the germ-plasm, and exposed as it is to the vicissitudes of an environment which is always hostile, it becomes at length an unfit envelope. But with the reproductive act the germ-plasm acquires a new soma, and it is no longer necessary that the former one should continue to exist as an unfit envelope. Physical death therefore occurs as an adaptation serving for the best interests of the race. The organism need not die, for the germ-plasm may be a physical continuum throughout innumerable generations. Somatic death is only a destructive metabolism : it is a catastrophic metabolism, if we like.

We may legitimately discuss such problems as the origin of the protoplasm of the prototrophic organism, or that of the chlorophyll-containing cell, or that of the nerve-cell. On the mechanistic view each of these conditions is a phase of a transforming physico-chemical system, and it is within the scope of the methods of physical science to investigate the nature of these transformations. But if the argument of this book is sound, then the problem of the origin of life, as it is usually stated, is only a pseudo-problem ; we may as usefully discuss the origin of the second law of thermodynamics ! If life is not only energy but also the direction and co-ordination of energies ; if it is a tendency of the same order, but of a different direction, from the tendency of inorganic processes, all that biology can usefully do is to inquire into the manner in which this tendency is manifested in material things and energy-transformations.   But the tendency itself is something elemental.

APPENDIX

INFINITY

WHAT is really meant when the mathematician uses the concept of infinity in his operations? Suppose that we take a line of finite length and divide it into halves, and then divide each half into halves, and so on *ad infinitum*. We make cuts in the line, and these cuts have no magnitude, so that the sum of the lengths into which we divide the line is equal to the length of the undivided line. We can divide the line into as many parts as we choose, that is, into an "infinite" number of parts.

Suppose that we are making a thing which is to match another thing, and suppose that we can make the thing as great as we choose. If, then, no matter how great we make the thing, it is still too small, the thing that we are trying to match is infinitely great.

Substitute "small" for "great," and this is also a definition of the infinitely small.

Clearly the idea of infinity does not reside in the *results* of an operation, but in its tendency. It inheres in our intuition of *striving* towards something, but not in the results of our striving.

---

[1] It must be understood that some of the things dealt with in these appendices are very hard to understand by the reader acquainted only with the results of biological science. We urge, however, that they are all relevant if biological results are to be employed speculatively.

FUNCTIONALITY

If we pour some mercury into a U-tube closed at one end, the air in this end will be contained in a closed vessel under pressure. We can increase the pressure by pouring more mercury into the open end of the tube. We can measure the volume of the air by measuring the length of the tube which it occupies. We can measure the pressure on this air by measuring the difference of length of the mercury in the two limbs of the tube. By taking all necessary precautions we shall find that for each value which the pressure attains there is a corresponding value of the volume of the air.

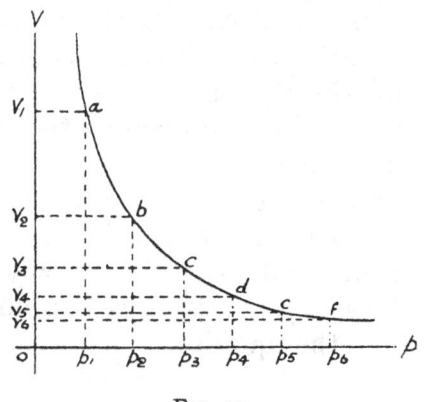

FIG. 27.

We thus find the pressure values, p_1, p_2, p_3, p_4, p_5, etc., and the corresponding volumes, v_1, v_2, v_3, v_4, v_5, etc., and we may then plot these values so as to make a graph.

In this figure the values represented along the horizontal axis are pressure-values, and those represented along the vertical axis are volume-values. We have so made the experiment that we can make the pressure-values whatever we choose—let us call them the values of the *independent variable* or *argument*. For each value of the pressure, or argument, there is a corresponding value of the volume, which *depends* on the pressure—let us call these values of the volume values of the *dependent variable* or *function*.

We can make arbitrary values of the pressure, but whenever we do this the corresponding values of the volume are fixed. We say, then, that the volume is a *function of the pressure*. In general, when we choose one value of an independent variable, or argument, there can be only one, or a small number, of values of the dependent variable, or function. If there are two or more values of the function for one value of the argument each of these is necessarily determined by the value which we choose to assign to the argument. There is a strict *functionality* between the two series of variables. In the experiment we have chosen this functionality is expressed by the equation $pv = k(1 + at)$, where p is the pressure, v the volume, k and a constants, and t is the temperature at which the experiment is carried out. In a number of experiments like that which we have mentioned, k, a, and t are the same throughout, and this is why we call them *constants*. We give p any value we like, and then v can be calculated from the equation.

RATE OF VARIATION

If we know the equation $pv = k(1 + at)$, we can find how much the volume changes when the pressure changes, that is, the rate of variation of v with respect to p. But even if we don't know that this equation applies, we can still find the rate of variation from our experiments. We see from the graph that, when the pressure increases from p_1 to p_2, the volume decreases from v_1 to v_2, but that if the pressure is again increased to p_3, that is, by a similar amount to the increase of pressure from p_1 to p_2, the volume decreases from v_2 to v_3. Now we find, by measurements made on the graph, that the decrease v_1 to v_2 is greater than the decrease v_2 to v_3, and the latter decrease is greater again than the decrease from v_3 to v_4. Evidently the rate of variation of volume is not like the rate of variation of pressure, that is, the same throughout, and when we look at the graph we see that the rate of variation is greatest where the slope of the curve is steepest. The latter is steepest near the point a, less steep near the point b, and still less steep near the point c. Now any *small* part of the curve is indistinguishable from a straight line. Let us draw a straight line ee_1, which appears to coincide with a small part of the curve near a, and similar straight lines ff_1, and gg_1, which also appear to coincide with small parts of the curve near b and c. Then the steepness of the curve will be proportional to the angles which these straight lines make with the axis op, and these angles are measured

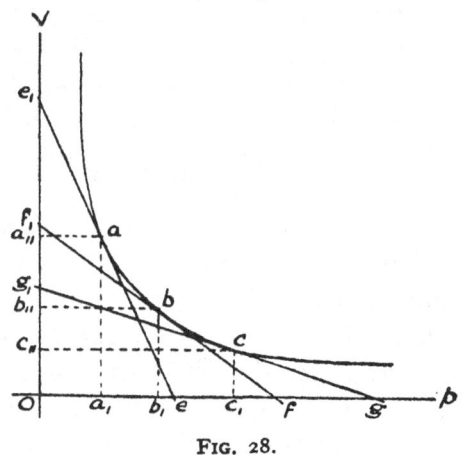

Fig. 28.

by their tangents, that is, by the ratio $\dfrac{oe_1}{oe}$ which is the tangent that e_1e makes with op, the ratio $\dfrac{of_1}{of}$, and the ratio $\dfrac{og_1}{og}$.

The point a on the curve corresponds with a pressure a_1 and a volume a_{11}. The point b corresponds with a pressure b_1 and a volume b_{11}, and c with a pressure c_1 and a volume c_{11}. The *average* rate of variation of the volume of the gas, as the pressure changes from a to c, is therefore proportional to the sum of the tangents $\frac{oe_1}{oe}$ and $\frac{og_1}{og}$, divided by 2.

Suppose that we wish to find the rate of variation of volume for a pressure change in the immediate vicinity of the value b_1, that is, the rate of variation as the pressure changes from a little less than b_1 to a little more than b_1. If we find the point b on the curve corresponding to b_1, and if we then draw a line ff_1, *touching* the curve at the point b, we shall obtain the angle off_1. It might appear now that the tangent of this angle, that is, the ratio $\frac{of_1}{of}$, would give us a measure of the rate of variation of volume.

But the reasoning would be faulty. The line ff_1 only *touches* the curve, it does not coincide with an element of the curve. Also at the point b_1 the pressure has a certain definite value, and there is no change. At the corresponding point b_{11} the volume also has a certain definite value, and there is no change. There can therefore be no rate of variation. The value of the tangent does not give us a measure of the rate of variation : it gives us the *limit* to the rate of variation, when the pressure is changing in the immediate vicinity of b_1.

We must stick to the notion of a pressure change in the *immediate vicinity* of b_1. What do we mean by " immediate vicinity " ? We mean that we are thinking of a range of pressure-values in which the particular pressure-value b_1 is contained, but not as an end-point. We mean also that we choose a definite standard of approximation to the value b_1, so that any pressure-value within our interval differs from b_1 by *less* than this standard of approximation. It means further that, no matter how small is the number representing this standard of approximation, *any* pressure-value within the interval will differ from b_1 by less than this number. This is what we really mean when we say that the interval we are thinking about is an " infinitely small one."

Now corresponding to this interval of pressure-values in the immediate vicinity of b_1, there will be an interval of volume-values in the immediate vicinity of b_{11}, and, as before, any one of these volume-values will differ from b_{11} by less than any number representing a standard of approximation to b_{11}. We then find the point on the curve corresponding to both b_1 and b_{11}, that is b, and we draw the line ff_1, and find the tangent of the angle which this line makes with op. The value of this tangent is the *limit* of the rate of variation of the volume of the gas when the pressure undergoes a change in the immediate vicinity of b_1.

" Rate of variation " is a function of the argument " pressure." This function has the limit l for a value of its argument b_1, when, as the argument varies in the immediate vicinity of b_1, the value of the function approximates to l within *any standard whatever* of approximation.[1]

We should not, of course, find the rate of variation of volume of the gas by this means. We should calculate the value of the differential co-efficient $\dfrac{dv}{dp}$ from the equation $pv=k(1+at)$: this would be $-k \dfrac{1+at}{p^2}$. But the reasoning involved in the methods of the calculus are those which we have attempted to outline.

---

[1] If the reader does not understand this, he should read Whitehead's "Introduction to Mathematics." He should read this book in any case.

We try to avoid the terms "infinitely small," "infinitely near," "infinitely small quantities," and so on, by the device of standards of approximation. It may appear to the non-mathematical reader that all this is rather to be regarded as "quibbling," but the success of the methods of mathematical physics should convince him that such is not the case. He should also reflect that clear and definite ideas on the fundamental concepts of the science are just as necessary in speculative biology as they are in mathematics.

(Another example.)

Let us consider the case of a stone falling from a state of rest. Observations will show that when the stone has fallen for one second it has traversed a space of 16 feet; at the end of two seconds it has fallen through 64 feet; and at the end of three seconds the space traversed is 144 feet. From these and similar data we can deduce the velocity of motion of the stone as it passes any point in its path.

The velocity is the space traversed in a certain time $\frac{s}{t}$. If we take any easily observable space (say five feet) on either side of the point chosen, and then determine the times when the stone was at the extremities of this interval, and divide the interval of space by the interval of time, we shall obtain the *average* velocity of motion of the stone over this fraction of the whole path chosen. But the velocity did not vary in a constant manner during this interval (as we see by considering the spaces traversed during the first three seconds of the fall). Therefore our average velocity does not accurately represent the velocity of the stone as it passes the point at the middle of the path chosen.

We therefore reduce the length of the path more and more so as to make the average velocity approximate closer and closer to the velocity near the middle portion of the path. In this way we find the ratio $\frac{\delta s}{\delta t}$, where δs is a very small interval of path containing the point chosen, but not as an end-point, and δt is a very small interval of time. Perhaps this average velocity may be near enough for our purposes, but perhaps it may not. The interval of path δs is still a finite interval, and δt is still a finite time, and so long as these values are finite ones the velocity deduced from them remains a mean one. All that we can say is that it approximates to the velocity, as the arbitrary point was passed, within a certain standard of approximation.

Obviously the smaller the interval δs, the closer will be this approximation. Suppose, then, that we diminish δs till it " becomes zero." It might appear now that when δs coincides with the point chosen we shall obtain the velocity of the stone at this point. But if there is no interval of path, and no interval of time, there can be no velocity, which is an interval of path divided by an interval of time ; and if the stone is " at the point," it does not move at all. We must stick to the idea of intervals of space and time, and yet we must think of these intervals as being so small that no error whatever is involved in regarding the mean velocity deduced from them as the " true velocity." We therefore think of the point as being placed in an interval of path, but not at an end-point of this interval. We think of the velocity as a mean one, but we must have a standard of approximation, so that we may be able to say that the mean velocity approximates to the " actual " or *limiting* velocity of the stone as it passes the point, within this standard of approximation. The smaller we make the interval, the closer will the mean velocity approximate to the limiting velocity.

We therefore think of the stone as moving in the immediate vicinity of the point in the sense already discussed. We say that the " immediate vicinity " is an interval such that any point in it, p_1, approximates to the arbitrary point p which we are considering within any standard of approximation : that is, no point in the interval is further away from p than a certain number expressing the standard of approximation, and this can be *any* number, however small. We say the same thing about the interval of time. That is to say, we make the intervals as small as we like : they can be smaller than any interval which will cause an error in our deduced velocity, no matter how small this error may be.

The limit of the velocity of a stone falling past a point in its path is, therefore, that velocity towards which the mean velocities approximate within any standard of approximation, when we regard the interval of space as being the immediate vicinity of the point, and the interval of time as being the time in the immediate vicinity of the moment when the stone passes the point. The limit of the velocity is not $\frac{\delta s}{\delta t}$ but $\frac{ds}{dt}$, dt and ds being, not finite intervals of time and space, but " differentials." We determine this limit by the methods of the differential calculus.

FREQUENCY DISTRIBUTIONS AND PROBABILITY

Let the reader keep a note of the number of trumps held by himself and partner in a large number of games of whist (the cards being cut for trump). In 200 hands he may get such results as the following :

No. of trumps in his own and partner's hands—o, 1, 2, 3, 4, 5, 6, 7, 8, 9, 10, 11, 12, 13.

No. of times this hand was held—o, o, o, 1, 9, 29, 53, 52, 35, 14, 6, 1, 0, 0.

He should note also the number of times that trumps were spades, clubs, diamonds, and hearts : he will get some such results as the following : spades, 46 ; clubs, 53 ; diamonds, 51 ; hearts, 50.

The numbers in the lower line of the first series form a " frequency distribution," for they tell us the frequency of occurrence of the hands indicated in the numbers above them. " No. of trumps " is the independent variable, and " no. of times these nos. of trumps were held " is the dependent variable.

A frequency distribution represents the way in which the results of a series of experiments differ from the mean result. A particular result is expected from the operation of one, or a few, main causes. But a number of other relatively unimportant causes lead to the deviation of a number of results from this mean or characteristic one. Yet since one, or a few, main causes are predominant, the majority of the results of the experiment will approximate closely to the mean ; and a relatively small proportion will deviate to variable distances on either side of the mean. If a pack of cards were shuffled so that all the suits were thoroughly mixed among each other, then we should expect the trumps to be as equally divided as possible between the four players. But a number of causes lead to irregularities in this desired uniform distribution, and so the results of a large number of deals deviate from the mean result: It is possible, by an application of the theory of probability, to calculate ideal, or theoretical frequency distributions, basing our reasoning on the considerations suggested above. We then find that the observed and calculated frequency distributions may be very much alike.

In biological investigation, far more than in physical investigation, we deal with mean results. It is, however, just as important that the mean should be considered as the individual divergences from the mean. We want to know the mean results, and the way and the extent in which the individual results diverge from the mean.

There is a mean or " ideal " result, but we must think of a great number of small independent causes which cause the actually obtained results to diverge from this mean. If these small un-co-ordinated causes are just as likely to cause the results to be less than the mean, as greater than the mean, we shall obtain a frequency distribution resembling the one given above, in that the variations from the mean are equal on both sides of the mean. But if the general tendency of the small un-co-ordinated causes is to cause the results, on the whole, to tend to be greater than the mean, then the frequency distribution will be " one-sided," that is, if we represent it by a curve the latter will be an asymmetrical one. Curves which are asymmetrical are those most frequently obtained in biological, statistical investigations.

MATTER

Our generalised notion of matter is that it is the physical substance underlying phenomena. Immediately, or intuitively, we attain the notion of matter because of our perceptions of touch, and our perception of muscular exertion. The distance sense-receptors, visual, auditory, and olfactory, would not give us this intuition of matter.

Material things are extended, that is, they have form, and they exclude each other, so that they cannot occupy the same place. They appear to us to be aggregates of different nature : they may be solid and homogeneous, like a piece of metal ; or solid and porous, like a piece of pumice-stone ; or loose and granular, like sand ; or viscous or liquid, like pitch or water. They may have colour. They are opaque, or transparent in various degrees. They may have odour. Material things, as they are perceived by the distance sense-receptors, appear to have qualities.

Material things are aggregates of molecules. The aggregates may possess essential form, like that of a crystal, or an organism. The form of the aggregate may be essential and homogeneous, so that it consists of molecules, all of which are of the same kind, like a crystal. It may be heterogeneous and essential, like the body of the organism, when it consists of molecules which are not all of the same kind. The aggregates may have accidental form, like that of a river valley, or a delta, or a mountain, and the form in these, and similar cases, is not a part of the essential nature of the aggregate.

The molecules are selections (in the mathematical sense) of some of about eighty different kinds of atoms. A molecule is a small number of atoms arranged to-gether in a definite way, and its nature depends, not only on the kinds of atoms of which it is composed, but also on the arrangement of these atoms. Two or more different arrangements of the same atoms are, in general, different molecules.

MASS

When matter is perceived by the tactile and muscular sense organs, we have the intuition of mass.

z

It is *heavy*, and the degree of heaviness is proportional to the quantity of matter in the body which we feel, that is, to its mass. Heaviness is synonymous with weight, but weight does not depend alone on the quantity of matter in the body. If the latter were removed to an infinite distance from the earth or other cosmic bodies, its weight would disappear, but its mass would remain. We could still touch and move it, and we should still find that different degrees of muscular exertion would be necessary when bodies of different masses had to be moved.

INERTIA

If the body were in motion, we should find that muscular exertion is necessary in order that it might be brought to rest; and if it were at rest, we should find that muscular exertion was necessary in order that it might be moved. The body, matter in general, possesses inertia, and this is its most fundamental attribute. Mass we can only conceive in terms of inertia. If two bodies were at rest, and if the same degree of muscular exertion conferred on each the same initial velocity of motion, their masses would be equal. If the same degree of muscular exertion conferred different velocities on different bodies, their masses would be different, and would vary directly with the initial velocities conferred.

FORCE

The feeling which we experience when we move a body from a state of rest, or stop a body which is moving, is what we call force. If on climbing a stair in the dark we think there is one step more than there is, and so have the queer, familiar, feeling of treading on nothing, we have the intuition of energy ; but when we tread on the steps, and so raise our body, we have the intuition of force. Force is that which accelerates the velocity of a mass. If the latter is at rest, we consider it to have zero velocity. If it is moving, and we stop it, there is still acceleration, but this is negative.

Matter, that is, the *substantia physica*, is clearly to be conceived only in terms of energy. It is, to our direct intuitions, resistance, or inertia, that which requires energy in order that it may be made to undergo change. Our static idea of physical solidity, or massiveness, disappears on ultimate analysis. Molecules are made up of atoms, and the atoms are assumed to have all the characters of matter : we could not *see* them, of course, even if we possessed all the magnifying power that we wished, for they would be too small to reflect light. Modern physical theory is compelled to regard atoms as complex, and imagines them as being composed of moving electrons. The electron is immaterial—it is the unit-charge of electricity. It is said to possess mass, but mass is now understood to mean inertia. So long as the electron is moving, it sets up a field of energy round it, and this field—the electro-magnetic one—extends in all directions. Periodic disturbances in it constitute radiation, and this radiation travels with the velocity of light. It is because of the existence of this field that we are *obliged* to postulate the existence of an ether of space. Unfamiliar to us until the discovery of Hertzian waves and "wireless" telegraphy, this electro-magnetic radiation in space is now accessible to our direct intuitions. We can initiate it by setting electrons in motion, that is, by expending energy (producing the sparking in the transmitters of the wireless telegraphy apparatus) ; and we can stop it, if it is in existence, by absorbing the energy (in the receivers of the wireless telegraphy apparatus). This is essentially what we understand by the inertia of gross matter. We set a body in motion by expending energy on it (the explosion of the powder in a cartridge, which converts potential chemical energy into the kinetic energy of the moving projectile) ; and we can stop a body which is in motion by absorbing this energy of motion (by causing the projectile to strike against a target, when the kinetic energy of its motion becomes the kinetic energy of the heat of the arrested body).

Inertia is therefore the same thing whether it be the inertia of visible, material bodies, or the inertia of invisible, material molecules, or the inertia of the immaterial, non-tangible ether. It is the condition that energy-changes must occur if anything accessible to our observation is to change its state of rest or motion.

ENERGY

Energy is therefore indefinable. It is an elemental aspect of our experience.

Nature to us is an aggregate of particles in motion. We have to speak of massive particles, whether we call these visible material bodies, or molecules, or atoms, or electrons, in order that we may describe nature. We must employ the fiction of a *substantia physica*. We only know the substance or matter in terms of energy ; it is really the latter that is known to us. It is the poverty of our language, or rather it is the legacy of a materialistic age, that compels us to speak of particles that move, rather than of motions as entities in themselves.

Considering, then, the idea of particles in motion as a fiction necessary for clear description, we can study energy. There is only one kind, or form, of energy which presents itself to our aided or unaided intuitions, that is kinetic energy. Bodies that move possess this energy represented by their motion : they can be made to do work, that is, their energy can be transformed into other forms of energy. All things are in motion. A gas consists of molecules incessantly moving with high velocity, and colliding and rebounding from each other. The energy of a gas is the sum of one-half of the masses of all the molecules, multiplied by the squares of the velocities of all the molecules, that is, $\Sigma\frac{1}{2}mv^2$. This is also the kinetic energy of a projectile, or of a planet revolving round the sun. Kinetic energy is that of the uniform, unchanging motion of some entity possessing mass, but we must extend our notion of mass so as to include immaterial, imponderable entities such as electrons.

This energy cannot be destroyed or created—the law of conservation of energy. This is a principle or mode of our thought. We are unable scientifically or philosophically to think of an entity ceasing to be. Dreams and phantoms show us entities which are real *while they last*, but which cease to exist. If we do attempt to think of entities that appear from, or disappear into, nothing, we surrender the notion of reality. The more we think of it the more clearly we shall see that the *things which we call real are the things which are conserved.*

Yet energy, to our immediate intuitions, seems to disappear. A flying bullet strikes against a target and becomes flattened out into a motionless piece of lead. A red-hot piece of iron cools down to the temperature of its surroundings. A golf-ball driven up the side of a hill comes to rest in the grass. A current of electricity passing through water is used up, that is, electricity of a higher potential is required to force the current through water than to force it through thick copper wire. In all these cases we might think that energy is lost, but we cannot believe this. The kinetic energy of the flying bullet becomes transformed into the increase of the kinetic energy of the molecules of the metal of which the bullet was composed; for the latter becomes greatly heated when its flight is arrested; and this increased heat ought to be equal to the kinetic energy of the bullet in flight. The red-hot piece of iron cools, and the kinetic energy of its molecules becomes less and less, but this does not cease to exist, for the energy is simply transferred by radiation and conduction to the surrounding bodies, the temperature of which it raises. The golf-ball driven up the hill comes to rest and loses its kinetic energy. Some of this has been transferred to the air through which it passes, the latter being heated very slightly; some of it is expended by friction with the grass over which the ball rolls before coming to rest, and this energy is traceable in heat-effects, or in mechanical effects, but the rest of it apparently ceases to exist. But this would be contradictory to the principle of conservation, and so we say that the lost kinetic energy has become potential. The current of electricity may heat the water through which it passes, and some of the energy which seems to disappear is so to be traced, but the greater fraction is apparently lost. A quantity of free hydrogen and oxygen is, however, generated, and we say that the kinetic energy of the moving electrons has become transformed into the potential chemical energy of the gaseous mixture.

POTENTIAL ENERGY

Therefore, if energy disappears or appears, we do not say that it is destroyed or is created : we invent potential energies, into which we suppose that the energies in question have become transformed, in order that we may still think of them as being subject to an *a priori* principle of conservation. Although a particle of radium continually generates heat, we do not therefore think of the first principle of energetics as being invalidated, for we suppose that the energy which thus appears was really potential in the atoms of radium. But it was contrary to all our former experience of atoms that they should contain any other energy than that of their own motion, and so the further assumption was made that the atom, at least the atom of the radioactive substance, is really complex, and not simple, as chemical theory demands. It is made up of smaller particles, and possesses a definite structure. In certain circumstances the atom may disintegrate, and the energy which held together its particles, whether these were simpler corpuscles or electrons, is given off as the heat which the radio-active substance apparently generates. The potential energy of the chemical atom is therefore a hypothesis which has been devised in order to preserve the validity of the law of conservation, and the reality of this hypothesis is being tested by investigation. If we accept it as true, are the deductions made from it justified in our experience ? That is the test which must be satisfied in all the hypotheses where potential energies are invented, and the potentials are only real if the test is satisfactory. The golf ball at rest at the top of the hill is a different entity from the golf ball at rest at the bottom of the hill : it is capable of developing energy, for a touch may cause it to roll down the hill, when most of the energy which was expended in order to drive it to the top of the hill will reappear in the form of the kinetic energy of motion of the ball. The atoms of hydrogen and oxygen which were dissociated by the energy of the electric current are different things from the atoms of hydrogen and oxygen which are combined together to form the molecules of water. Their state when the gases are in the elementary condition, or are " free," is that of molecules moving rapidly and incessantly, rebounding from each other after colliding with each other : they possess energy of position—potential energy—because they are separate from each other. If they " combine," as when a minute electric spark explodes the mixture of gases, they tractate together, and remain in proximity to each other, becoming molecules of water. The energy which became potential in the gaseous mixture, when the electric energy of the current seemed to disappear, now appears as the heat generated by the combustion, that is, as the greatly increased kinetic energy of the molecules of the gas (steam) which takes the place of the mixture of hydrogen and oxygen. Previous to the explosion this gas was a mixture of molecules of hydrogen and oxygen ($2H_2+2O$) at the ordinary temperature, but after the explosion it consists of a smaller number of molecules at a very much higher temperature.

What is " energy of position " ? The golf ball at the bottom of the hill was at a distance of R feet from the centre of the earth, but at the top of the hill it is at a distance of $R+100$ feet from the centre of the earth. In the first case it was free to fall R feet, but in the second case it is free to fall $R+100$ feet. The atoms of the constituent molecules of water occupy the position $H - O - H$, the bonds $(-)$ indicating that the atoms are very close together; but when the water is decomposed by an electric current, the atoms occupy the positions $O - O + H - H + H - H$, the $(+)$ indicating that the atoms are relatively

far apart from each other. Now the golf ball and
the earth, or the atoms of hydrogen and oxygen,
are physically the same material entities, whether
they are close together or far apart, yet when
the earth and the ball, or the atoms of oxygen and
hydrogen, are separated from each other, their " properties " are different from what they are when they are
close together. What is it that makes the difference ?
It is that which is *between* them. Is it, in the last case,
" the potential energy of chemical affinity " ? This
dreadful phrase is actually used in a recent book on
biology : " In the elements carbon and oxygen, so
long as they remain separate, a certain amount of
energy remains latent. When the carbon and oxygen
atoms are allowed to come together and unite, this
potential energy of chemical affinity is liberated as
kinetic energy." What is changed by the tractation
and pellation (the terms suggested by Soddy in place
of the anthropomorphic ones, " attraction " and " re-
pulsion ") ? It is the ether which has become changed
in some way. Potential energy resides therefore in
the ether of space.

ISOTHERMAL AND ADIABATIC CHANGES

Let us consider the changes which occur in a gas
under the influence of changes in temperature and
pressure, premising that the remarks which we have
to make can be applied to bodies in the liquid and
solid conditions, with some necessary modifications.
A gas, then, consists of a very great number of particles,
or molecules, in motion. These molecules move in
straight lines at very high velocities, and if the envelope
in which the gas is contained is a restricted one, the
molecules collide with each other, and with the walls

of the envelope; and, being assumed perfectly elastic, they rebound from each other, and from the walls of the vessel, with the same velocity which they had when they collided. The pressure of the gas (say that of steam at a temperature of 110° C., and a pressure of 120 lbs. to the square inch in a steam boiler) is the sum of the impacts of the molecules on the walls of the containing vessel. When the temperature is high the molecules are moving at a higher mean velocity than when the temperature is lower, and their mean free path tends to become greater. The volume of a certain mass of gas, that is, the volume occupied by a certain very great number of molecules, is greater the higher is the temperature, provided the envelope is one capable of yielding. If we reduce the capacity of the envelope in which the gas is contained, the pressure will rise, for the intrinsic

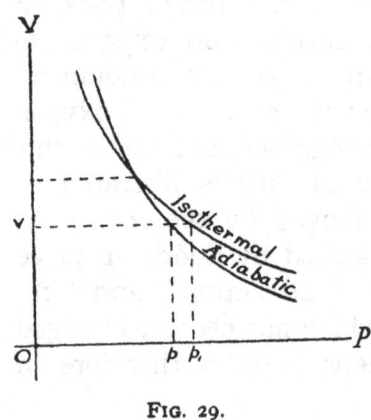

FIG. 29.

energy of the gas is still the same; but we have done work on it, and by the law of conservation this work, or at least the energy represented by it, must still exist. It is represented by the decreased length of free path of the molecules, and this means that the impacts on the walls of the vessel will be greater than they were. There is, therefore, a certain relation between the volume of a gas and its pressure, and this relation can be represented by an equation involving the temperature, the pressure, and the volume.

The diagram represents the pressure and the volume of a gas when these things change. There are two conditions, (1) when the heat developed by the compression is allowed to escape through the walls of the vessel to the outside, or when the heat lost in the expansion of the gas is compensated by the conduction of heat through the walls of the vessel from outside ; and (2) when the heat developed is retained in the gas, as when the latter is contained in a vessel the walls of which do not conduct heat. The pressure of the gas is measured along the horizontal axis, and the volume is measured along the vertical axis, and a curve is drawn so that for any value of the pressure there is a corresponding value of the volume. Thus the values of the pressures p and p_1 in the diagram correspond to the value of the volume v. The curve relating the change of pressure with a corresponding change of volume is, in general, that called a rectangular hyperbola. But there are two kinds of such curves : (1) that which we obtain by plotting the corresponding values of pressure and volume, when the temperature of the gas remains constant throughout the series of changes, that is, when the rise of temperature which would occur when the gas is compressed is compensated by the conduction of this heat to the outside of the vessel containing the gas. Such a series of changes of pressure and volume is called an *isothermal* one. (2) When the heat developed by the compression of the gas is retained in the gas, as when the walls of the vessel in which these changes are effected are such as do not conduct heat : such a series of changes is called an adiabatic one. Adiabatic curves are steeper than are isothermal ones.

THE CARNOT ENGINE

This is an imaginary mechanism which performs a certain cycle of operations. It does not really exist, but the conception of its operation is of the greatest value in the consideration of energy-transformations, and it is for this reason that we discuss it here.

Consider a gas, or some other substance capable of expanding or contracting. It contains intrinsic energy, and it is capable of doing work. Thus, since a gas can expand indefinitely it can be made to do mechanical work. A mass of gas at a pressure p_1, and having a volume v_1, and at a temperature $T°$, can do work by expanding till its pressure is reduced to p, and its volume increased to v. If it expands adiabatically its temperature will fall to $t°$. Let us suppose that $t°$ is the temperature of the surrounding medium : the gas cannot therefore

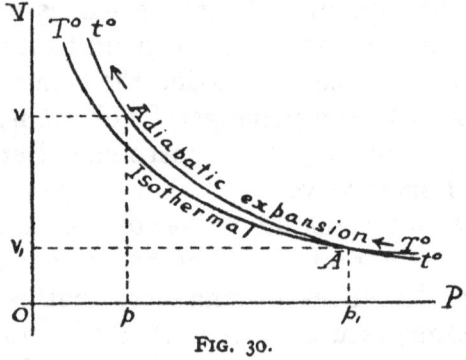

FIG. 30.

cool further, and we can obtain no more work from it. If the gas is the substance which we wish to employ as the working substance in the Carnot engine, we must therefore bring it back to the condition represented by A. That is, we must raise its temperature to $T°$, we must reduce its volume to v_1, and we must increase its pressure to p_1.

Thus the steam of an engine is (say) at a temperature of 110° C., and a pressure of 120 lbs. to the square inch. When it has passed through the cylinder and condenser it is water at a temperature of, say, 15° C., and it is at atmospheric pressure. We must, therefore, bring it back to its former condition by heating this water in the boiler till it is steam under the former conditions of temperature and pressure.

Therefore we must, in order to obtain a self-acting engine, cause the working substance, and the mechanism of the engine, to perform a series of cyclical operations.

The Carnot engine is a cylinder containing a gas called the working substance S, and this gas can be brought into thermal contact with a source of heat, or a refrigerator, that is, the gas can be heated or cooled by a mechanism outside itself. The walls of the cylinder are made of some substance which is a perfect non-conductor of heat, but the bottom of the cylinder is made of a substance which conducts heat perfectly. There is a piston in the cylinder which fits it closely, but which moves up and down without friction. At the bottom of the latter is a valve which can be turned so as to place the bottom of the cylinder, and therefore the gas, in thermal contact with a reservoir of heat $(+)$, or a refrigerator $(-)$.

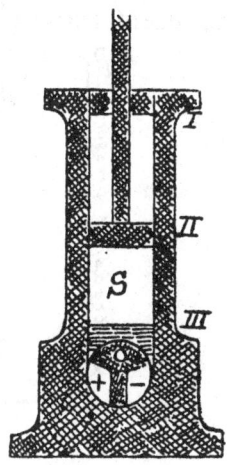

Fig. 31.

But when the valve is turned so that the non-conducting part O fills the bottom, the gas is perfectly insulated, and heat can neither enter nor leave it.

Such an engine is, of course, an imaginary one, since there can be no mechanism in which there is not a certain amount of friction between moving parts, and there are no substances which conduct or insulate heat perfectly. The engine is, in fact, the *limit* to a series of engines each of which is supposed to be more perfect than the last one. It is a fiction which is of considerable use in theoretical work.

We have therefore a substance which can be heated by contact with a hot body, and which can then expand, doing mechanical work by raising a piston, and perhaps turning a flywheel, and on which work is then done so that it returns to its original condition. This is a cycle of operations. If we consider only the changes which occur in the working substance we can represent these changes by a diagram.

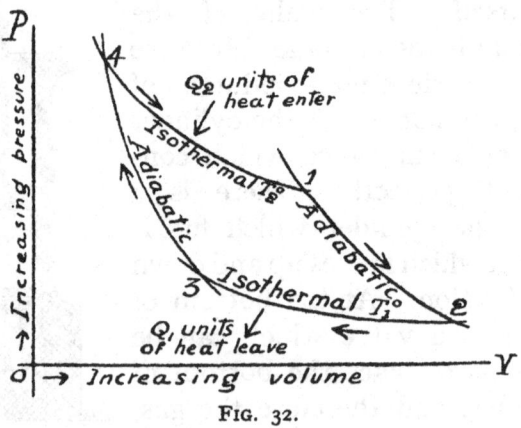

FIG. 32.

First operation, (1→2). We suppose that the valve is turned so that the non-conducting plug closes the cylinder. The piston is in the position II (Fig. 31). Heat cannot then enter or leave the gas. But the latter already contains heat : it is at a temperature of $T_2°$, so that it can expand doing work. Let it expand, forcing up the piston. During this operation the pressure of the gas will fall from a point on the vertical axis opposite 1 to a point opposite 2, and its volume will increase from a point on the horizontal axis beneath 1 to a point beneath 2. It will cool because it has expanded, and no heat is allowed to enter it during this act of expansion. The expansion is therefore adiabatic; the temperature falls from $T_2°$ to $T_1°$; and work is done *by* the gas.

Second operation, (2→3). The piston is now at the position I, that is, at the upper end of its stroke, and we must bring it back again to the lower end of the cylinder. The valve is turned so that the bottom of the cylinder is placed in thermal communication with the refrigerator ($-$), and the piston is pushed in to the position II. The gas is therefore compressed until its volume decreases from a point beneath 2 to a point beneath 3. As it is being compressed, heat is generated and its temperature would rise, but as this heat is generated it flows into the refrigerator, so that the temperature of the gas remains the same during the operation. The contraction is therefore an isothermal one; the temperature remains at $T_1°$; and work is done *on* the gas from outside.

Third operation, (3→4). But the piston is not at the lower end of its stroke yet. We turn the valve so that the bottom of the cylinder is closed by the non-conducting plug O, and then push in the piston until it reaches the position III. The gas is still further compressed, and this compression generates heat. But the heat cannot escape, so that the temperature of the gas rises until it reaches $T_2°$. The contraction is therefore an adiabatic one. Work is done *on* the gas.

Fourth operation, (4→1). The piston is now at the lower end of its stroke. We turn the valve so that the bottom of the cylinder is placed in communication with the source of heat ($+$). The gas expands from the point beneath 4 to the point beneath 1, raising the piston to the position II. This expansion of the gas would lower its temperature, but it is in communication with the source of heat, and so it does not cool, but draws heat from the source and remains at a constant temperature, $T_2°$. The expansion is therefore an isothermal one. Work is done *by* the gas.

This completes the cycle. But the gas is heated, and when the piston is at position II, the valve is turned so as to close the cylinder by the non-conducting plug O. The heat already contained in the gas continues to expand, the latter doing more work, but this expansion causes the temperature to fall from $T_2°$ to $T_1°$. This is the operation with which the cycle commenced.

Summarising the positive Carnot cycle, we see that the engine takes heat from a source (+) and gives up part of this to a refrigerator (-), (in an actual steam-engine heat is taken from the boiler and given up to the condenser water). If we measure the quantity of heat taken from the boiler in the steam which enters the cylinders we shall find that this quantity of heat is greater than the quantity which is given up to the condenser water. What becomes of the balance? It is converted into the mechanical work of the engine. The Carnot engine therefore takes a quantity of heat, Q_2, from the source and gives up another quantity of heat, Q_1, to the refrigerator. We find that Q_2 is greater than Q_1, and the balance, $Q_2 - Q_1$ is represented by the work done by the engine. Heat-energy falls from a state of high, to a state of low potential, and is partly transformed into mechanical work.

THE CARNOT NEGATIVE CYCLE

This is simply the positive cycle *reversed*. The reader should puzzle it out for himself if he is not already familiar with it. It consists of an adiabatic contraction 2→1, an isothermal contraction 1→4, an adiabatic expansion 4→3, and an isothermal expansion 3→2. A quantity of heat, Q_1, is taken from the refrigerator at a temperature $T_1°$, and another quantity, Q_2, is given up to the source at a temperature $T_2°$. But Q_2 is greater than Q_1, and the engine therefore gives up more heat than it receives, while, further, heat flows from a body at a low temperature to another body at a higher temperature. Where does the engine get this energy from? It gets it because work is done *upon* it by means of an outside agency, and all of this work is converted into heat.

REVERSIBILITY

The Carnot engine and cycle are therefore perfectly reversible. Not only can the engine turn heat into work, but it can turn work into heat. This perfect, quantitative reversibility is, however, a property of the imaginary mechanism only, and it does not exist in any actual engine.

ENTROPY

Let us consider the cycle more closely. In the operation 4→1, which is an isothermal expansion, there is a flow of heat-energy from the source and a transformation of energy into work. The gas in the condition represented by the point 4 had a certain pressure and a certain volume. In the condition represented by the point 1, its pressure has decreased, its volume has increased, and its temperature is the same. Its physical condition has been changed, and to bring it back into its former condition something must be done to it. Let, then, the gas continue to expand without receiving any more heat, or parting with any: that is, let it

2 A

undergo the adiabatic expansion 1→2 until its temperature falls to that of the refrigerator, $T_1°$. We now compress the gas while keeping it at this temperature, that is, we cause it to undergo the isothermal contraction 2→3, during which operation it is giving up heat to the refrigerator, so that there is again a flow of heat-energy. We then compress it still further without allowing heat to escape from it, that is, we cause it to undergo the adiabatic contraction 3→4. During this operation the gas rises in temperature to $T_2°$. It is now in the condition that it was when the cycle commenced.

In this cycle of operations heat first entered, and then left the gas, and with this entrance or rejection of heat, the condition of the gas with respect to its power of doing work changed. We investigate this flow of heat, and the concomitant change of properties of the substance, with regard to which the flow took place, by forming the concept called *entropy*. We make the convention that when heat enters a substance the entropy of the latter increases, and when heat leaves it its entropy decreases. We call the quantity of heat entering or leaving a substance Q, and the temperature of the substance T. Then $\frac{Q}{T}$ is proportional to the change of entropy of the substance when the quantity of heat, Q, enters or leaves it.

Now it is a fact of our experience that heat can only flow, *of itself*, from a hotter to a colder body. Consider two such bodies forming an isolated system, the temperature of the hotter one being $T_2°$, and that of the colder one $T_1°$. Let Q units of heat flow from the body at $T_2°$ to that at $T_1°$ no work being done. Then the loss of entropy of the hotter body is $\frac{Q}{T_2°}$, and the gain of entropy of the colder body is $\dfrac{Q}{T_1{}^\circ}$. The nett change of entropy of the system is $\dfrac{Q}{T_1{}^\circ} - \dfrac{Q}{T_2{}^\circ}$. Since $T_2{}^\circ$ is greater than $T_1{}^\circ$, $\dfrac{Q}{T_2{}^\circ}$ is less than $\dfrac{Q}{T_1{}^\circ}$. There-fore the expression $\dfrac{Q}{T_1{}^\circ} - \dfrac{Q}{T_2{}^\circ}$ is positive, that is, the entropy of the system, as a whole, has increased. When heat flows from a hotter to a colder body the nett entropy of the two bodies, therefore, in-creases.

But we can also cause heat to flow from a colder to a hotter body *by effecting a compensatory energy-trans-formation.* Such a compensation would not occur *by itself* in any system capable of effecting an energy-transformation, if it is to be effected some external agency must act on the transforming system. We can suppose it to happen in a perfectly reversible imaginary mechanism. Suppose a Carnot engine works in the positive direction, taking heat from a reservoir at temperature $T_2{}^\circ$, and giving up part of this heat to a refrigerator at $T_1{}^\circ$, and doing a certain amount of work W. Suppose that this work is stored up, so to speak, say by raising a heavy weight, which can then fall and actuate the same Carnot engine in the opposite (negative) direction. The engine then exactly reverses its former series of operations. The work it did is reconverted into heat, and as much of this heat flows from the refrigerator into the source, that is, from a colder to a hotter body, in the negative operations, as flowed from the source to the refrigerator in the positive operations. In this primary energy-transfor-mation, combined with a compensatory energy-trans-formation, there is no change of entropy. The mechanism is an ideal one—the limit to an irreversible mechanism.

But—and now we appeal to experience and cease to work with ideal mechanisms—the actual engine which we can design and work is one in which there will be friction, in which some parts will conduct heat imperfectly, and other parts will insulate heat imperfectly. Let the friction generate q units of heat, and let the quantity of heat which is " wasted " by imperfect conduction and insulation be q_1. This heat will flow into the refrigerator, or will be radiated or conducted to the surrounding medium, which we suppose to be at the same temperature as the refrigerator. If, then, we divide this total quantity of heat by the temperature $T_1°$, we get $\dfrac{q+q_1}{T_1°}=S_1$ as the quantity of entropy which is generated as the result of the imperfections of the engine, in addition to the quantity of entropy, S, which would be generated if the engine were a perfect one. Both S and S_1 are positive.

Also in the working of the engine in the negative direction a certain quantity of entropy, S_1, is generated for reasons similar to those mentioned above.

The entropy generated when the engine works in the positive direction is therefore $S+S_1$, and when it works negatively the quantity generated is also S_1. The entropy destroyed when the engine works negatively is S. The total change of entropy is therefore $2S_1+S - S$, that is, $2S_1$. In an actual energy-transformation combined with a compensatory energy-transformation there is therefore an increase of entropy.

We can generalise these statements so that they will apply not only to a heat-engine but to all mechanisms which effect energy-transformations. In all such transformations entropy is generated. Therefore *the Entropy of the Universe tends to a maximum.*

AVAILABLE AND UNAVAILABLE ENERGY

Consider the Carnot engine as a perfect mechanism. It takes heat-energy from a source at a temperature $T_2°$, and it gives up heat to a refrigerator at a temperature $T_1°$, $T_2°$ being greater than $T_1°$. In the adiabatic expansion $1 \to 2$ the gas continues to expand until its temperature becomes equal to that of the refrigerator. It cannot, then, expand and do work any longer; and thus the proportion of the heat, Q_2, received from the source, which can be converted into work, depends on the difference of temperature $T_2° - T_1°$. The greater is this difference the greater will be the proportion of the heat-energy received which can be converted into work. If the engine were a perfect one, and if the gas were also a perfect one (that is a gas which would continue to expand according to the equation for the adiabatic expansion of gases), and if the refrigerator were absolutely cold, then *all* the heat energy received from the source could be converted into work.

We cannot produce a refrigerator of absolute temperature $0°$, and therefore only a certain proportion of the heat which is received by the engine can be transformed into mechanical work. But this work can be used to reverse the action of the engine, and thus the same fraction of the total heat-energy which was given to the refrigerator can be taken from it and given back to the source. The perfect engine is therefore reversible without loss of available energy.

Now consider still the engine as a mechanism which takes heat from a source and gives it to a refrigerator, but let it be an actual engine. Instead of giving up

a certain fraction of the heat received to the refriger-
ator—a fraction equal to $Q_1 \frac{T_1^{\circ}}{T_2^{\circ}}$, it gives up rather
more, because it is not a perfect mechanism, that is, it
generates friction, etc. Some of the heat received
thus ceases to be available for the performance of
work; and passes into the refrigerator. The fraction
of the heat-energy which passes into the refrigerator
in the perfectly reversible engine was unavailable
energy in the conditions in which the mechanism
worked, or was imagined to work, but in the actual
engine this fraction is increased. If we divide the
increase of unavailable energy by the temperature of
the refrigerator, the product is the increase of entropy
generated in the actual engine over that generated
in the ideal engine. Because of this reduction of
available energy the actual engine is an irreversible
mechanism.

This is the connection between unavailable energy
and entropy. In all transformations some fraction
of the transforming energy becomes heat, and this
heat flows by conduction and radiation into the surrounding bodies. In general this heat simply raises
the temperature of the medium into which it flows,
and becomes unavailable for further transformations.
With every transformation that occurs some part of
the energy involved becomes unavailable. Therefore
although the sum of the available and unavailable
energy of the Universe remains constant, the fraction of
unavailable energy tends continually to a maximum.

INERT MATTER

We can see now what is indicated by Bergson's
" inert matter." It is not matter deprived of energy

—such an expression has no meaning—*it is energy which is unavailable for further transformations.*

The matter in which we choose to say that this energy is inherent has become *inert.* Let us substitute for the Carnot engine the actual steam-engine of a ship, the condenser of which is cooled by the sea water which is taken in, and which is then heated and flows out again into the sea. The heat derived from the source, that is, from the furnace of the boiler where coal is burned to raise steam, thus passes out into the sea. Now the heat capacity of the sea is so great that the temperature of the water is not appreciably raised by this heat, which drains into it from the engine : even if it were appreciably raised, the heat would be conducted into the earth, or would be radiated out into space, and would then raise the temperature of the material bodies of the universe. But let all this heat remain in the sea. It then simply raises the temperature of the water by an exceedingly small amount, and the motions of the molecules become infinitesimally increased. But the heat becomes equally distributed by conduction and convection throughout the mass of the water in the sea, and as there are no differences in adjacent parts there are no means whereby the energy which thus passes into the sea can be again transformed.

A new order of things is the result of the processes we have indicated. The segregated, available heat-energy of material bodies has become transferred to the un-co-ordinated, diffuse, unavailable energies of the molecules which compose these bodies. The transformations which we can effect depend on the condition that the energy which we utilise is that of aggregates of molecules which are in a different physical condition, as regards this energy, from adjacent aggregates. But when this energy becomes equally distributed among the molecules of all the aggregates, the matter in which it inheres becomes inert. If we could, by a sorting process like that of Maxwell's hypothetical demons, a process which does not expend the energy with which it deals, separate the molecules which were moving slowly from those which were moving more quickly, we could make this energy again available. But it must clearly be understood that our physics is the physics not of individual molecules, but of aggregates of molecules.

INDEX

Variation, rate of (mathematical), 344 ;
in biology, 186 ;
atavistic, 195 ; direction of, 233 ; fluctuating, 189 ; must be co-ordinated, 231 ; mathematical probability of co-ordination of, 233 ; the material for selection, 229 ; origin of, 230 ; selected by the organism, 237 ; cause of, a pseudo-problem, 242 ; arise de novo, 244.
Variables (mathematical), 343.
Varieties, specific, 194.
Vegetable life, 265.
Vertebrates, 249 ;
adaptations securing mobility, 275 ; ancestry of, 253 ; morphology of, 249 ; a dominant group, 259 ; distribution of, 260.
Verworn, and mechanism in life, 127.
Vesalius, anatomical school of, 121.
Vital activities, integration of, 128 ; co-ordination of, 171
de Vries and mutations, 191 ;
fluctuating variations inherited, 220.
Vital force, 318.

Van der Waal's equation, 308.
Weber's law, 16 ;
a quasi-mathematical relation, 17.
Weismann, hypothesis of heredity, 182 ;
hypothesis of germinal selection, 241 ; hypothesis of development, 132 ; mosaic-theory, 131 ; preformation hypothesis, 133 ; hypothesis of the germ-plasm, continuity of the germ-plasm, 181 ; germinal changes inconceivable, 224 ; size of biophors, 183 ; origin of life, 339 ; spontaneous generation a logical necessity, 339.
Weismannism, a series of logical hypotheses, 320 ; physico-chemical analogies, and subsidiary hypotheses, 223.
Whales, an unsuccessful line of evolution, 274.
Whitehead, and mathematical reasoning, 347.
Wilson, mosaic-theory of development, 139.

Yerkes, and behaviour of crustacea, 293.

Zymogens, 92.
Zymoids, 94

Printed in the United States
By Bookmasters